T0214706

Synthesis Lectures on Operations Research and Applications

This series focuses on the use of advanced analytics in both industry and scientific research to advance the quality of decisions and processes. Written by international experts, modern applications and methodologies are utilized to help researchers and students alike to improve their use of analytics. Classical and cutting-edge topics are presented and explored with a focus on utilization and application across a range in practical situations.

Bruce Golden · Xingyin Wang · Edward Wasil

The Evolution of the Vehicle Routing Problem

A Survey of VRP Research and Practice from 2005 to 2022

Bruce Golden
Robert H. Smith School of Business
University of Maryland
College Park, MD, USA

Xingyin Wang
Workforce Optimizer Pte Ltd.
Singapore, Singapore

Edward Wasil
Kogod School of Business
American University
Washington, DC, USA

ISSN 2770-6303   ISSN 2770-6311   (electronic)
Synthesis Lectures on Operations Research and Applications
ISBN 978-3-031-18718-6   ISBN 978-3-031-18716-2   (eBook)
https://doi.org/10.1007/978-3-031-18716-2

This Springer imprint is published by the registered company Springer Nature Switzerland AG
The registered company address is: Gewerbestrasse 11, 6330 Cham, Switzerland

Preface

The Vehicle Routing Problem (VRP) is one of the most widely studied problems in combinatorial optimization. This is due to its mathematical complexity and practical significance. The challenges and advancements in VRP research and practice from 2005 to 2014 have been detailed in three major books by Golden, Raghavan, and Wasil [1], Toth and Vigo [2], and Corberán and Laporte [3]. The chapters in those three books described technological advancements such as global positioning systems and radio frequency identification and their impact on the VRP, as well as techniques for modeling VRP variants and solving large-scale problems more quickly and accurately.

It has been nearly a decade since the publication of the most recent book and the field of VRP research and practice has grown by leaps and bounds. The number of new variants, solution methodologies, and application areas has exploded. The time is right for an updated source on this topic that provides valuable coverage of the large number of recent survey articles that fully span the wide range of areas in vehicle routing.

In this book, we identify 135 articles published in scholarly, academic journals from January 2005 to June 2022 that survey various aspects of the VRP ranging from exact and heuristic solution methods to new problem variants such as drone routing to new research areas such as green routing. We catalog and classify these articles, make key observations about publication history and overall contributions, and identify trends in VRP research and practice such as last-mile delivery, urban distribution, and dynamic and stochastic routing.

We hope that readers find our book interesting, informative, and useful. We trust that it will serve as a valuable resource to researchers and practitioners with ongoing or unfolding research efforts regarding the VRP for many years to come.

College Park, MD, USA
Singapore, Singapore
Washington, DC, USA
August 2022

Bruce Golden
Xingyin Wang
Edward Wasil

v

References

1. B. GOLDEN, S. RAGHAVAN, and E. WASIL, *The Vehicle Routing Problem: Latest Advances and New Challenges*. Springer, New York, 2008.
2. P. TOTH and D. VIGO, *Vehicle Routing: Problems, Methods, and Applications, Second Edition*. MOS-SIAM, Philadelphia, 2014.
3. Á. CORBERÁN and G. LAPORTE, *Arc Routing: Problems, Methods, and Applications*. MOS-SIAM, Philadelphia, 2014.

Contents

The Evolution of the Vehicle Routing Problem—A Survey of VRP Research and Practice from 2005 to 2022

1

1.1 Introduction

Since its introduction in the late 1950s, the Vehicle Routing Problem (VRP) has been a very significant area of research and practice in operations research. Searching on the words vehicle routing problem in Google Scholar (August 25, 2021) produced about 546,000 results—an impressive number of results, but time-consuming to cull through if you are looking for published work in a specific area such as inventory routing (which has about 114,000 results). Fortunately, over the years, researchers and practitioners have written survey articles that compile published work on models, algorithms, and applications for a specific problem usually over a 5- to 10-year span. A survey article is a valuable and efficient way to learn about what has been accomplished and what needs to be done (future work) in a specific area of VRP research and practice. Note that we will use the terms survey article and article interchangeably in the rest of this book.

Recently, we noticed that certain new areas of interest, such as the routing of drones, had already produced several survey articles. We wondered what the survey landscape looked like for the entire VRP field going back 15 years or so. To keep the research and practice recent and relevant, we conducted a search from January 2005 to June 2022 and identified 135 survey articles published in scholarly, academic journals (we did not consider survey articles appearing in books or in conference proceedings). We relied heavily on Google Scholar and Google Scholar Alerts (received daily by one of us, dating back to 2011). For each survey article identified, we searched backwards via the list of references and forward via the list of subsequent articles citing the survey, in order to identify other survey articles. In the remainder of this book, we catalog and classify these articles, make key observations about publication history, summarize the overall contributions of each article, and identify trends in VRP research and practice.

© The Author(s), under exclusive license to Springer Nature Switzerland AG 2023
B. Golden et al., *The Evolution of the Vehicle Routing Problem*, Synthesis Lectures on Operations Research and Applications,
https://doi.org/10.1007/978-3-031-18716-2_1

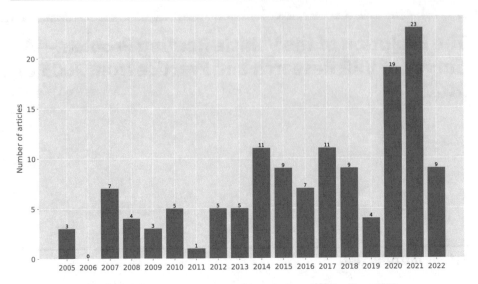

Fig. 1.1 Number of VRP survey articles published from January 2005 to June 2022

In Fig. 1.1, we show the number of survey articles published in each year. The number of articles starts to increase in 2014, with more than 30% occurring in 2020 (19 articles) and 2021 (23 articles). In Table 1.1, we list the top 10 journals by number of survey articles published (note that the 135 articles appear in 65 journals that include some far outside mainstream operations research outlets; for example, two articles appear in Cogent Engineering), and the total citation counts for these articles. These journals have published 67 articles (49%) with a total of 21,764 citations. The European Journal of Operational Research has published the most articles (17), about 13% of the total, followed by Computers & Operations Research (8), Computers & Industrial Engineering (8), Transportation Science (7), and Networks (7).

Although a simple listing of the references for the 135 survey articles might be somewhat useful to those conducting research in an area (for example, a graduate student deciding on a dissertation topic), we decided to provide additional detail about each article. The additional detail would help to guide and focus a search for published work in a specific area of VRP research and practice.

We developed four categories for the type of survey article: overview, models, solution methods, and applications. Models were broken down into 15 categories ranging from articles that covered alternative and multiple objectives to articles that covered specific VRP variants. In Table 1.2, we show the 18 tables that we developed to categorize the 135 articles. Three articles appear in more than one table. Articles covering green routing (16), applications (15), overviews (13), exact algorithms and heuristics (11), and dynamic and stochastic routing (10) occur most often, about 48% of the total.

Table 1.1 Popular journals for VRP survey articles

Journal	Number of articles	Total citation counts	Average citation count per article
European Journal of Operational Research	17	8002	471
Computers & Industrial Engineering	8	2655	332
Computers & Operations Research	8	1671	209
Networks	7	1327	190
Transportation Science	7	5392	770
EURO Journal on Transportation and Logistics	6	752	125
International Transactions in Operational Research	5	464	93
Annals of Operations Research	3	1019	340
Journal of the Operational Research Society	3	405	135
Yugoslav Journal of Operations Research	3	77	26
Total	67	21764	

In the next section, we provide detailed comments about the contribution of each article to VRP research and practice.

1.2 Overall Contributions of Survey Articles

In this section, for each article in the 18 tables, we provide detailed comments on its contributions to the literature including a count of the number of references in the article (this gives an indication of the depth of an article's overall coverage of a VRP topic).

Table 1.2 Categories of VRP survey articles

Table	Title	Number of articles
1.3	Overview	13
1.4	Alternative and Multiple Objectives	3
1.5	Arc Routing and General Routing	5
1.6	Drones, Last-mile Delivery, and Urban Distribution	8
1.7	Dynamic and Stochastic Routing	10
1.8	Green Routing	16
1.9	Inventory Routing	7
1.10	Loading Constraints	5
1.11	Location-Routing	5
1.12	Multiple Depots	4
1.13	Pickup and Delivery and Dial-a-Ride Problems	8
1.14	Rich and Multi-attribute Routing	4
1.15	Routing Over Time	8
1.16	Shipping	3
1.17	Two-echelon, Collaborative, and Inter-terminal Problems	5
1.18	Specific Variants, Benchmark Datasets, and Software	9
1.19	Exact Algorithms and Heuristics	11
1.20	Applications	15

Table 1.3 Overview

Article	Year	Reference count	Comments
Tan and Yeh [122]	2021	153	Review of recent VRP research (2019–2021)
			88 articles were selected
			15 large tables are provided
			Detailed taxonomy
			VRP variants
			Model features
			Model objectives
			Solution approaches
			Observations and conclusions
Zhang et al. [134]	2021	186	Review of the VRP
			Basic VRP
			Solution procedures (four exact algorithms, 13 heuristics)
			Mathematical model of the capacitated VRP and descriptions of solution procedures including ant colony optimization
			Mathematical model of VRP with time windows and descriptions of solution procedures
			Variants of VRP with time windows (simultaneous pickup and delivery, multiple trips with fuzzy demands, fuzzy time windows, service choice, multiple depots)
			Mathematical model of the split delivery VRP
			Analysis of feasible solutions
			Exact algorithms and heuristics including heuristics for time windows
			Definition and classification of the dynamic VRP
			Route update strategy
			Optimization algorithms and heuristics
			VRP variants including simultaneous pickup and delivery, backhauls, and multiple trips
			Research trends including parallel algorithms, design of standard test cases, and multi-objective green VRP with mixed constraints

(continued)

Table 1.3 (continued)

Article	Year	Reference count	Comments
Konstantakopoulos et al. [63]	2020	89	Review and classification of the VRP for logistics distribution
			Detailed discussion of variants faced by most logistics and distribution companies including capacitated, open, multi-trip, time windows, consistent, and time dependent problems
			Identified 263 articles related to freight transportation that were published from 2010 until the first quarter of 2020
			Classified exact, heuristic, and metaheuristic algorithms for solving variants
			Identified trends in variants (e.g., capacitated VRP is the most studied) and in algorithms (e.g., metaheuristics dominate the field) over the years
			Highlighted the most significant publications on 16 variants ranging from the capacitated VRP to time windows, to pickups and deliveries, to the green VRP
Vidal et al. [130]	2020	323	Review of vehicle routing variants
			Objectives, performance metrics, or combination of objectives including the consideration of
			Profitability
			Service quality
			Equity
			Consistency
			Simplicity
			Externalities
			Integration with other decisions such as
			Districting
			Facility location
			Fleet composition
			Inventory and production management
			More fine-grained models include
			Specificities of the transportation network
			Specificities of drivers and vehicles
			Specificities of customer requests
			Challenges and prospects

(continued)

Table 1.3 (continued)

Article	Year	Reference count	Comments

Jayarathna et al. [54]	2019	34	Short review of major progress in the VRP
			History
			Practical impact
			Complexity, NP-hard
			Formulations
			Variants: time windows, multi-depot
			Solution approaches
			    Exact methods, set partitioning
			    Heuristics
			    Metaheuristics

Jayarathna et al. [56]	2019	164	Review of recent model developments and improvements in the VRP
			Mathematical formulations
			    Two-index and three-index flow formulations for the capacitated VRP
			    Time windows
			    Multiple depots
			Exact solution methods
			    Algorithms based on the set partitioning formulation
			Heuristic, metaheuristic, and hybrid solution approaches
			Uncapacitated and capacitated facility location problems
			Mathematical formulation for the single source, capacitated, multi-facility location problem
			Facility location models: median, center, and covering problems
			Definition and mathematical formulation of the location-routing problem

Goel and Maini [46]	2017	79	Survey of solution methods for the VRP from 2009 to 2016 Descriptions of six variants: time windows, pickup and delivery, time dependent, backhauls, dynamic, and stochastic Descriptions of solution methods (exact, heuristic, metaheuristic) with a summary of the methods used in 29 papers

(continued)

Table 1.3 (continued)

Article	Year	Reference count	Comments
Braekers et al. [16]	2016	309	Classification and review of 277 VRP articles published from 2009 to 2015
			Count of articles published by journal and classification of articles by year of publication
			Proposed taxonomy with five main characteristics from type of study to data characteristics
			Overview of five solution methods such as exact method and classical heuristic
			Overview of 16 problem characteristics including time windows, backhauls, and multiple depots
			Detailed summaries of three variants: open VRP, dynamic VRP, and time-dependent VRP
Eksioglu et al. [34]	2009	96	Classification of VRP literature
			Brief history of VRP literature
			VRP definition and background
			Count of 1021 articles published by year from 1954 to 2006 and by academic journal
			Proposed taxonomy with five main characteristics from type of study to data characteristics
			Test of proposed taxonomy on 30 articles
Laporte [70]	2009	89	Commemorates 50th anniversary of the VRP (1959 to 2009)
			VRP definition
			Exact algorithms (branch and bound, dynamic programming, and vehicle flow, commodity flow, and set partitioning formulations and algorithms)
			Classical heuristics (savings, set partitioning, cluster-first and route-second, and improvement)
			Metaheuristics (local search, population search, learning mechanisms, and computational results)

(continued)

Table 1.3 (continued)

Article	Year	Reference count	Comments
Liong et al. [74]	2008	51	VRP models and solution methods
			Classical definition
			Mathematical formulations: capacitated VRP, VRP with time windows, and VRP with pickups and deliveries
			Algorithms (exact methods such as column generation and heuristics)
Laporte [69]	2007	89	Summarizes results for the classical VRP with capacity constraints
			Exact algorithms (two-index flow, two-index two-commodity flow, and set partitioning formulations)
			Classical heuristics (constructive and improvement)
			Metaheuristics (local search, population search, learning mechanisms, and computational comparison of metaheuristics for the classical VRP from 17 papers published between 1993 and 2007)
Marinakis and Migdalas [78]	2007	206	Annotated bibliography
			Capacitated VRP
			Seven metaheuristics including simulated annealing, tabu search, and genetic algorithms
			VRP with time windows
			Seven metaheuristics including simulated annealing, tabu search, and genetic algorithms
			Variants including pickups and deliveries, backhauls, multiple depots, multiple periods, and heterogeneous fleet

Table 1.4 Alternative and multiple objectives

Article	Year	Reference count	Comments
Rossit et al. [116]	2019	86	Visual attractiveness in vehicle routing
			Importance of visually attractive solutions
			Literature review of articles addressing visual attractiveness in routing and districting problems
			Different ways of assessing visual attractiveness of routes
			Nine compactness measures
			Two overlap and crossings measures
			Three complexity measures
			Computational experiments
			Correlation analysis of the different attractive measures
			Visual comparison included in the appendix
			Recommendations
Matl et al. [81]	2018	80	Workload equity in vehicle routing
			Review of the vehicle routing literature that considers equity
			Vehicle routing with route balancing (VRPRB)
			Time window extensions to VRPRB
			Min-max vehicle routing problem
			Other vehicle routing variants with equity aspects
			Vehicle routing applications with equity aspects
			Theoretical analysis
			Eight axioms for desirable inequality measures including inequality relevance, transitivity, scale invariance, translation invariance, population independence, anonymity, monotonicity, and Pigou-Dalton transfer principle
			Six commonly used inequality measures including min-max, lexicographic min-max, range, mean absolute deviation, standard deviation, Gini coefficient
			Insights from the theoretical analysis
			Numerical analysis
			All feasible solutions to 60 small instances with 14 customers are generated to find the Pareto-optimal solutions
			General observation on poor equity of cost-optimal solutions, small marginal cost of equity, and mixed agreement between different equity functions
			Drawbacks of non-monotonic equity measures
			Future research directions

(continued)

Table 1.4 (continued)

Article	Year	Reference count	Comments
Jozefowiez et al. [58]	2008	70	Multi-objective vehicle routing problems Introduce the multi-objective vehicle routing Definition of dominance and potentially Pareto-optimal solution relative to an algorithm Motivation and examples of multi-objective vehicle routing Extend classic academic problems Generalize classic problems Real-life applications Common objectives considered Algorithms for multi-objective optimization

Table 1.5 Arc Routing and General Routing

Article	Year	Reference count	Comments
De Maio et al. [29]	2021	131	Literature review on arc routing under uncertainty Arc routing background Stochastic and robust arc routing problems Stochastic programming Robust optimization Stochastic programs with recourse Chance-constrained programs Uncertainty theory, fuzzy logic, and other frameworks
Wang and Wasil [131]	2021	79	50 years of vehicle routing papers in Networks Arc routing problem 11 articles summarized
Mourão and Pinto [89]	2017	210	Updated annotated bibliography on arc routing (2010 - 2017) Single vehicle arc routing problems Multiple vehicle arc routing problems Arc routing applications
Baldacci et al. [8]	2010	42	Generalized VRP Provides a uniformed framework for various VRP extensions TSP with profits VRP with selective backhauls Covering VRP Capacitated general windy routing problem Routing Automated Guided Vehicles (AGVs)
Corberán and Prins [26]	2010	150	Annotated bibliography on arc routing Uncapacitated arc routing problems Capacitated arc routing problems

Table 1.6 Drones, Last-mile Delivery, and Urban Distribution

Article	Year	Reference count	Comments
Liang and Luo [72]	2022	83	Truck-drone routing problem (TDRP)
			More than 60 papers since 2015 are examined
			Two base models described and formulated
			TSP with drone (MILP)
			VRP with drone (0-1 ILP)
			Incorporate flight endurance and drone capacity into the model
			Detailed classification of 64 TDRP articles
			According to five attributes: number of trucks, features related to drones (e.g., capacity), cooperative modes between trucks and drones, objective (e.g., minimize completion time), and customer requirements (e.g., time windows)
			Overview of exact algorithms in 11 articles including dynamic programming and branch-and-bound
			Overview of heuristics in four categories: operators-based algorithms (15 articles); evolutionary algorithms (11 articles); reassigned algorithms and Markov decision processes (9 articles); decomposition and model-based algorithms (5 articles)
			Discussion of extending research in six areas including dynamic orders, environmental conditions, and regulations on drone operations
Na et al. [90]	2022	117	Last-mile logistics (LML)
			Comprehensive review of studies relevant to LML
			Definitions of LML terms including last-mile distribution
			Four current issues in LML
			Sharing economy
			Proximity stations, points and hubs
			Routing problem for last-mile delivery hubs
			Location-routing model
			Hybrid genetic algorithm
			Two-stage stochastic travel time model
			Environmentally sustainable LML
			New opportunities for delivery services
			Development of optimization methods for delivery routing problems that must be addressed in a short time period (e.g., VRP using drones)
			Develop algorithms and heuristics to solve real-world problems that may be large-scaled

(continued)

Table 1.6 (continued)

Article	Year	Reference count	Comments
Boysen et al. [15]	2021	273	Last-mile delivery
			Challenges in last-mile delivery including increasing volume and time pressure
			Classification scheme for storage, transportation, and handover
			Current delivery concepts with a discussion of the setup of infrastructure, staffing and fleet size, and routing and scheduling
			Human-driven delivery vans including routing and scheduling with time windows and time-dependent travel times
			Cargo bikes
			Customer self-service with handover of a parcel at a decentralized facility such as a parcel locker or shop including routing and scheduling
			Near-future delivery concepts with a discussion of the setup of infrastructure, staffing and fleet size, and routing and scheduling
			Drones
			Autonomous delivery robots
			Crowdshipping
			Use of free capacity of transport options dedicated to moving people
			Alternative handover options
			Farther-future delivery options
			Alternative drone-launching platforms
			Autonomous driving
			Tunnel-based cargo transport
			Future research
Macrina et al. [76]	2020	145	Review of drone-aided routing
			Cover papers published between 2015 and May 2020
			Review 63 articles that focus on routing problems for parcel delivery
			Classification based on: main areas of application (civilian, environment, and defense), drone size, shape of fuselage (e.g., fixed wing), type of propulsion (e.g., gas turbine, electric, and battery)
			Four categories of routing problems: TSP with drones including the flying sidekick TSP and variants, VRP with drones, drone delivery problem, carrier-vehicle problem with drones
			Mathematical models, algorithms, and instances for each of the four categories
			Directions for future research
			Environmental impacts
			Energy evaluation
			Consideration of more realistic drone parameters such as type of fuselage
			Use of drones in a dynamic delivery system

(continued)

Table 1.6 (continued)

Article	Year	Reference count	Comments
Rojas Viloria et al. [114]	2020	95	Review of unmanned aerial vehicle routing problems
			Contributions between 2005 and 2019
			Descriptive analysis of 79 papers
			Number of papers published per year and number appearing in main journals
			Papers with single applications including parcel delivery and surveillance/data collection and multiple applications including humanitarian logistics and parcel delivery
			Optimization objectives in 79 papers (50 optimize one objective, 29 are multiobjective)
			Exact methods such as branch-and-bound used by 48 papers, heuristics such as nearest neighbor used by 54 papers, and metaheuristics such as tabu search used by 30 papers
			Models with ground vehicles to support drones during the routing (25 papers)
			Modeling constraints such as time windows and battery restrictions (addressed in 39 papers)
			Trends and future directions (e.g., humanitarian approaches and last-mile deliveries)
Thibbotuwawa et al. [123]	2020	142	Review of unmanned aerial vehicle routing problems
			Different types of routing characteristics encountered in unmanned aerial VRP including time windows, multiple depots, and split deliveries
			Routing with land-based and maritime-based modes of transportation
			Degree of automation
			Mathematical formulation of the unmanned aerial VRP
			Current state of research in the field including top seven subject areas and publication trends in top journals and conferences
			Overview of VRP and TSP approaches used in 18 and seven papers, respectively, on the unmanned aerial VRP
			Challenges in routing unmanned aerial vehicles (weather, air traffic control, fuel consumption and range)

(continued)

Table 1.6 (continued)

Article	Year	Reference count	Comments
Otto et al. [96]	2018	321	Optimization problems in operations planning of drones
			Civil applications of drones including physical infrastructure, agriculture, and transport
			Planning drone operations including area of coverage, planning of search operations, and routing for a set of locations
			Planning combined operations of drones with other vehicles including vehicles supporting operations of drones, drones supporting operations of vehicles, and drones and vehicles synchronized as working units
			Strategic, tactical, and operational issues including integration of drones into civil airspace and location of physical infrastructure
			Model extensions and new problems motivated by drone research
			Future research directions including dealing with uncertainty and drone design
Cattaruzza et al. [22]	2017	113	Vehicle routing for distribution in cities
			Literature review of vehicle routing for city logistics including VRP with access time windows
			Classification of urban goods movement (inter-establishment, end consumer, urban management)
			Statistical analysis of urban goods movement flows
			Main challenges in the optimization of vehicle routes for urban goods movements
			Time-dependent travel times
			Data collection and management
			Multi-level distribution
			Dynamics of the city
			Dynamic travel and service times
			Routes with multiple trips

Table 1.7 Dynamic and Stochastic Routing

Article	Year	Reference count	Comments
Ojeda Rios et al. [94]	2021	146	Review of the dynamic vehicle routing problem from 2015 to 2021
			Taxonomy of the problem
			Taxonomy of the solution methods
			Dynamic and deterministic problems
			Dynamic and stochastic problems
			Solution methods
			New business-to-consumer applications
			Future work and opportunities
Soeffker et al. [119]	2021	112	Review of the stochastic dynamic vehicle routing problem (SDVRP) in the light of prescriptive analytics
			Main sources of uncertainty (demands, environment, resources)
			Information models and decision models
			Computational methods to derive policies
			How to approach a specific SDVRP with prescriptive analytics
Sangeetha and Srinivasan [117]	2020	52	Review of static, dynamic, and stochastic vehicle routing in home healthcare
			Home healthcare VRP
			Patient information
			Objective function
			Constraints (e.g., matching skills to patients, synchronization, time windows)
			Planning horizon
			Solution methods
Ulmer et al. [125]	2020	84	Literature review on modeling stochastic dynamic VRPs
			More than half the papers have appeared since 2010
			Authors focus on route-based Markov decision process models
			Review compares and contrasts modeling techniques and solution methods
			Establishes a basis for combining route-based optimization with dynamic and stochastic modeling
			Using examples from the literature, the authors show that their models closely align with cutting edge solution approaches

(continued)

Table 1.7 (continued)

Article	Year	Reference count	Comments
Oyola et al. [98]	2018	85	Literature review on stochastic VRP models
			Roughly covers 1996-2016
			Types of stochasticity
			Most common type—VRP with stochastic demands
			Evaluation of solutions
			Common recourse actions
Oyola et al. [97]	2017	92	Literature review on stochastic VRP solution methods
			Roughly covers 1996–2016
			Exact methods (e.g., integer L-shaped method)
			Heuristic approaches (e.g., adaptive large neighborhood search)
Psaraftis et al. [110]	2016	169	Dynamic vehicle routing
			Dynamic VRP taxonomy
			Based on 11 criteria
			Focus on trends since 2000
Ritzinger et al. [113]	2016	124	Survey on dynamic and stochastic VRPs
			Dynamic VRPs
			Stochastic VRPs
			Dynamic and stochastic VRPs
			Dynamic VRP with stochastic travel times
			Dynamic VRP with stochastic demands
			Dynamic VRP with stochastic customers
			Dynamic VRP with multiple stochastic aspects
			Production and routing interface (e.g., the inventory routing problem)
Berhan et al. [12]	2014	84	Survey on the stochastic VRP
			VRP and variants
			Stochastic VRP
			49 stochastic VRP articles are summarized
Pillac et al. [105]	2013	154	Review of dynamic VRPs
			Information evolution and quality
			Dynamic VRPs
			Applications
			Solution methods

Table 1.8 Green Routing

Article	Year	Reference count	Comments
Asghari and Al-e-hashem [7]	2021	329	Review of the green VRP
			Three major research streams
			Internal combustion engine vehicles
			Alternative-fuel powered vehicles
			Hybrid electric vehicles
			Combining green VRP with traditional VRP features such as time windows and fleet size and mix
			Solution methodologies (exact, heuristic, metaheuristic)
			Application areas
			Main insights, trends, and future research directions (e.g., genetic algorithms, adaptive large scale neighborhood search, and variable neighborhood search are popular solution methods)
Dündar et al. [33]	2021	213	Review of sustainable urban vehicle routing
			Review of papers published from 1975 to 2019 on routing problems with sustainability concerns
			Studies and findings on economic, environmental, and social dimensions
			Modeling effort including data type (e.g., real data) and solution methods (exact, heuristic, simulation)
Kucukoglu et al. [67]	2021	15	Review of the electric vehicle routing problem (EVRP)
			Classification of 136 research articles based on
			Objective function
			Energy consumption calculation
			Constraints
			Fleet type
			Mathematical model for the EVRP and some variants
			Solution approaches: Exact and heuristic methods
			Benchmark datasets
			Future research directions
Marrekchi et al. [79]	2021	135	Recent advances in the green VRP
			Systematic review of studies on the green VRP from 2014 to 2019
			Negative externalities such as noise pollution and congestion
			Fuel consumption models
			Taxonomy of studies
			Optimization methods used in 86 papers
			Classification of papers on nature of the data (stochastic or deterministic), nature of the fleet, objective function, constraints)
			Benchmark instances
			Future directions

(continued)

Table 1.8 (continued)

Article	Year	Reference count	Comments
Moghdani et al. [84]	2021	161	Review of the green VRP
			Systematic review of 309 papers from 2006 to 2019
			Classification of VRPs including pollution routing, electric vehicles, time dependent, and energy minimizing
			Solution methodologies (exact methods, exact solvers, metaheuristics)
			Future research directions and opportunities (e.g., single-objective and multiple-objective optimization approaches)
Normasari and Lathifah [92]	2021	100	Heterogeneous fleet green VRP
			Review of 90 papers from 2009 to 2020 that involve the green VRP with some papers covering heterogeneous fleet variants such as routing with time windows
			Green VRP with a single objective and multiple objectives
			Solution approaches (exact, heuristic, metaheuristic, hybrid)
Qin et al. [111]	2021	116	Electric vehicle routing problems
			Comprehensive review of routing problems that use electric vehicles and consider recharging operations on the routes
			Nine classes of problems are described including problem settings and methods of solution
			Electric TSP
			Green VRP (solution approaches used in eight papers from 2012 to 2019 are described)
			Electric VRP (mixed integer formulation; description of variants and solution approaches in 23 papers from 2014 to 2020)
			Mixed electric VRP (features and solution approaches in four papers from 2015 to 2019)
			Electric location routing problem (features and solution approaches in five papers from 2015 to 2019)
			Hybrid electric VRP
			Electric dial-a-ride problem
			Electric two-echelon VRP
			Electric pickup and delivery problem with time windows

(continued)

Table 1.8 (continued)

Article	Year	Reference count	Comments
Xiao et al. [132]	2021	79	Electric vehicle routing problem
			Systematic review
			Charge station visit model
			Charging time optimization
			Electricity/energy consumption model
			Practical factors
			Charging
			Time windows
			Mixed fleet
			Solution approaches
			Exact
			Traditional heuristic
			Metaheuristic
			Hybrid heuristic
			Summary of critical factors in modeling including load, travel speed, and battery swapping
			Energy consumption rate modeling for an electric vehicle
			New formulation of charging station visiting and recharging
			New comprehensive model of the electric vehicle routing problem with time windows
			Computational experiments with the comprehensive model on 29 instances accounting for the effects of air conditioning and load, battery capacity, and travel speed
Alfaseeh and Farooq [1]	2020	55	Three-factor taxonomy of eco-routing models
			Aggregation level of traffic flow and emissions/fuel models
			Scalability (small case studies, large case studies)
			Number of objectives optimized
			Potential research directions
			Multi-objective routing
			Predicting traffic characteristics such as speed, flow, and density

(continued)

Table 1.8 (continued)

Article	Year	Reference count	Comments
Ferreira et al. [39]	2020	87	Multi-objective optimization for the green VRP
			Review of 76 papers from 2012 to 2018
			Detailed summary of papers on
			Application type (case study, general application)
			Fleet type
			Environmental considerations
			Triple bottom line requirements (economic, environmental, social)
			Solution procedure (exact, heuristic) and specific techniques
			Number of objectives
			Description of application areas
			Number of citations
Erdelić and Carić [37]	2019	175	Survey of the electric VRP
			Overview of electric VRP variants and related problems in 80 papers (detailed summary table)
			Variants of the electric VRP including the electric shortest path problem, electric TSP, mixed vehicle fleet, hybrid vehicles, partial recharging, and green and pollution routing problem
			Solution procedures (exact, heuristic, metaheuristic) for the electric VRP in 79 papers (detailed summary table)
			Future research directions
Srivatsa Srinivas and Gajanand [120]	2017	69	Review of driver behavior and the VRP
			Features of the VRP with distance/cost minimization, time minimization, and fuel consumption and pollution minimization including fleet types, time windows, and solution approaches
			Studies on driver behavior
			Non-congested networks (driving speed)
			Congested networks (route choice)
			Integrating vehicle routing and driver behavior models
Toro et al. [124]	2016	123	Review of the VRP in the context of green transportation
			Classification of routing problems including the emissions VRP
			Formulations for the VRP based on vehicle flow, commodity flow, and set partitioning
			Solution techniques (exact, heuristic, metaheuristic, matheuristic)
			Emission estimation techniques
			Summary of 12 papers that model fuel consumption rate
			Trends and future directions in the green VRP

(continued)

Table 1.8 (continued)

Article	Year	Reference count	Comments
Demir et al. [30]	2014	118	Research on green road freight transportation
			Factors affecting fuel consumption including vehicle weight and speed
			Fuel consumption models (13 macroscopic models, 12 microscopic models)
			Fuel consumption models for road transportation planning
			Applications of 10 models
			Descriptive statistics for 58 papers including publication type, solution approach, and type of emission model
			Five areas for future research including integration of microscopic models within GIS software and factors that may reduce emissions such as location of the depot and selection of the right vehicles
Lin et al. [73]	2014	287	Survey of the green VRP
			Review of 19 VRP variants including split deliveries, time windows, and multiple depots
			Algorithms and benchmark instances
			Green VRP
			Studies from 2007 to 2012
			Pollution-routing problem
			Studies from 2007 to 2012
			VRP in reverse logistics
			Studies in four areas including waste collection and end-of-life goods collection
			Trends and future directions in the green VRP including more operational constraints in waste collection and multi-echelon distribution system
Park and Chae [101]	2014	42	Review of solution approaches for the green VRP
			Fuel consumption models
			Formulation of capacitated VRP with fuel consumption rate
			Exact solution methods including branch-and-bound and dynamic programming
			Heuristic solution methods including route construction and neighborhood search
			Summary of 23 studies that use metaheuristics including tabu search and genetic algorithms
			Summary of 14 solution approaches to the green VRP

Table 1.9 Inventory Routing

Article	Year	Reference count	Comments
Shaabani [118]	2022	113	Review of the perishable inventory routing problem (PIRP)
			89 relevant papers
			Perishable products
			52 different journals and conference proceedings
			Distribution of PIRP papers by journal
			Five classification attributes
			Type of product
			Number (single, multiple)
			Type of demand
			Objective function
			Solution approach
			Breakdown of papers by attributes
			Type of perishable products studied
			Distribution of solution approaches
			Different demand types
			PIRP papers with multiple objectives
			40 papers with a case study
			Papers published by year
			Conclusion and future work
Roldán et al. [115]	2017	66	Inventory routing with stochastic demands and lead times
			Information management and inventory policies
			Stochastic demand and lead time modeling
			Optimization methods
			Multiple depot inventory routing problem
Coelho et al. [24]	2014	139	Inventory routing
			Classification schemes according to problem structure and information availability
			Application of inventory routing
			Basic version of the inventory-routing problem (static and deterministic): Exact and heuristic algorithms
			Extension of the basic inventory-routing including the production-routing problem, problem with multiple products, problem with direct deliveries and transshipment, and consistent inventory-routing problem
			Stochastic inventory routing
			Dynamic inventory routing
			Benchmark instances

(continued)

Table 1.9 (continued)

Article	Year	Reference count	Comments
Bertazzi and Speranza [14]	2013	70	Inventory routing
			Introduce inventory routing with multiple customers
			Single-vehicle inventory routing: Mathematical formulation, valid inequalities, and properties
			Multi-vehicle inventory routing: Mathematical formulation and properties
			Literature review of papers on inventory routing from 1985 to 2013 organized by the objective functions
Bertazzi and Speranza [13]	2012	35	Inventory routing
			Introduce the main characteristics of inventory routing: Planning horizon, structured policies, objectives, and decisions
			Analytic solutions and worst-case analyses for the single link shipping problem (one customer) with continuous time
			Mathematical formulations for the single link shipping problem with discrete time
			Worst-case analyses for the inventory routing problem with direct shipping (multiple customers)
			Quick literature review of the pioneering papers in the 1980s and other papers from 1990 to 2011 on single link and direct shipping problems, and surveys and tutorials on inventory routing
Andersson et al. [2]	2010	125	Inventory routing
			Industrial aspects of inventory management and routing
			Type of supply chains
			Conditions for beneficial integration
			Current practice
			Difference between road-based and maritime transportation
			Research aspects of inventory management and routing
			Separate description of inventory management and routing
			Key components of the combined problem, including time, demand, topology, routing, inventory, and fleet size and composition
			Literature review of more than 90 articles from 1983 to 2009 including papers on industrial applications
			Trends and future directions
Moin and Salhi [85]	2007	49	Inventory routing
			Single-period models
			Multi-period models
			Infinite horizon models
			Stochastic models
			Research directions

Table 1.10 Loading Constraints

Article	Year	Reference count	Comments
Sun et al. [121]	2021	78	Distribution of finished automobiles
			Overview of the automobile shipping problem
			Decision makers
			Service models
			Transportation models
			Decision makers
			Introducing the automobile shipping optimization problem
			Automobile shipping optimization problem at the operational level
			Loading and reloading optimization without routing: Formulations, algorithms, and test instances
			History of routing literature with loading consideration
			Optimization models considering both loading and routing: Objective functions, constraints, algorithms, and test instances
			Automobile shipping by rail and sea
			Automobile shipping optimization problem at the strategic and tactical levels
			Mode selection and network design
			Empty autorack management
			Research direction including hybrid service model, full integration of loading and routing, and realistic routing constraints
Ostermeier et al. [95]	2021	90	Muli-compartment vehicle routing problems
			Mathematical formulation for a general multi-compartment vehicle routing problem
			Attributes of the multi-compartment vehicle routing problem
			Compartment-related attributes such as flexibility in compartment size, assignment of product types to compartments, and shareability of compartments
			Order fulfillment-related attributes such as split delivery
			Other VRP attributes not exclusive to multi-compartment problem
			Literature review of 84 publications from 1981 to 2020 grouped into application areas
			Fuel distribution
			Waste collection
			Agricultural transportation
			Grocery distribution
			Maritime transportation
			Other application including bike-sharing system, chemical products distribution, and passenger transportation
			Conceptual problems without specific applications
			Discussion
			Attributes and applications
			Methodology and applications
			Future research directions

(continued)

Table 1.10 (continued)

Article	Year	Reference count	Comments
Pollaris et al. [106]	2015	110	Survey of vehicle routing problems with loading constraints
			VRP characteristics that influence loading and routing including vehicle and cargo characteristics, time-dependent travel duration, legal requirement on the maximum driving time, customer request types, and multiple objectives
			Loading constraints
			Multi-dimensional packing constraints
			Cargo-related constraints including complete-shipment constraints, allocation constraints (connectivity and separation), and positioning constraints
			Container-related constraints including weight limits, weight distribution
			Item-related constraints including priorities, orthogonality, orientation, and stacking constraints
			Stability constraints
			Integration of loading and routing
			Two-dimensional loading capacitated vehicle
			Three-dimensional loading capacitated vehicle routing problem
			Multi-pile vehicle routing problem
			Multi-compartment vehicle routing problem
			Pallet packing vehicle routing problem
			Minimum multiple trip vehicle routing problem with incompatible commodities
			Traveling salesman problem with pickups and deliveries with LIFO/FIFO constraints
			Double traveling salesman problem with pickups and deliveries with multiple stacks
			Vehicle routing problem with pickups and deliveries with additional loading constraints
			Benchmark instances
			Research directions

(continued)

Table 1.10 (continued)

Article	Year	Reference count	Comments
Iori and Martello [53]	2013	79	Annotated bibliography on combined routing and loading problems
			Vehicle routing
			Loading
			Two-dimensional packing
			Three-dimensional packing
			Real-world loading constraints
			Pickup and delivery with loading
Iori and Martello [52]	2010	93	Review of routing problems with loading constraints
			Brief introduction to routing problems
			Brief introduction to loading constraints
			Capacitated vehicle routing problem with two-dimensional loading constraints
			Exact and metaheuristic approaches
			Computational comparison of the metaheuristics
			Capacitated vehicle routing problem with three-dimensional loading constraints
			Metaheuristic approaches
			Computational comparison of the metaheuristics
			Multi-pile vehicle routing problem
			Exact and metaheuristic approaches
			Computational comparison of the metaheuristics
			Traveling salesman problems with pickup and delivery and loading constraints; with LIFO loading constraints; with FIFO loading constraints
			Double traveling salesman problem with multiple stacks
			Miscellaneous problems including VRPs with multi-compartment loading

Table 1.11 Location-Routing

Article	Year	Reference count	Comments
Mara et al. [77]	2021	305	Review of 222 articles
			English language journals
			Published between 2014 and 2019
			Provides detailed taxonomy of location-routing problems (LRPs)
			Charts LRP publication growth from 2007 to 2019
			Shows top journals for papers
			Indicates types of studies
			Classifies 66 case studies
			Catalogs solution approaches
			Displays scenario characteristics
			Lists the physical characteristics of LRP papers
			Identifies the objective functions in LRP papers
			Presents future research directions
Jayarathna et al. [56]	2019	164	Vehicle routing problem
			Capacitated vehicle routing problem and its mathematical formulations
			Capacitated vehicle routing problem with time windows and its mathematical formulations
			Vehicle routing problem with multi-depot and its mathematical formulations
			Heuristic and exact solution methods
			Discrete facility location problem
			Brief review of the uncapacitated facility location problem, capacitated facility location problem, and single source capacitated (multi) facility location problem
			Mathematical formulations for the single source capacitated (multi) facility location problem, median problem, center problem, and covering problem
			Locational routing problems
Prodhon and Prins [108]	2014	116	Location-routing problem
			Survey of 72 articles published from 2007 to 2013
			Capacitated and uncapacitated location-routing problems (exact and heuristic methods, approximation algorithms, and benchmark instances)
			Multi-echelon location-routing (two-echelon, mobile depots, and truck and trailer routing)
			Problems with special or multiple objective functions
			Problems with additional attributes on nodes and vehicles, with multiple periods, with inventory management, and with uncertain data
Lopes et al. [75]	2013	165	Location-routing problem
			Two-level taxonomy (hierarchical structures and algorithmic approaches) of 149 articles

(continued)

Table 1.11 (continued)

Article	Year	Reference count	Comments
Nagy and Salhi [91]	2007	144	Location-routing problem
			Exact methods
			Clustering-based, iterative, and hierarchical heuristics
			Stochastic location routing problems
			Dynamic (multi-period) location routing problems
			Problems with non-standard hierarchical structures

Table 1.12 Multiple Depots

Article	Year	Reference count	Comments
Jayarathna et al. [55]	2021	26	Survey of the VRP with multiple depots
			Mathematical formulation
			Summary of 12 papers from 2011 to 2019 that use exact solution methods
			Summary of 12 papers from 2010 to 2020 that use heuristic and metaheuristic solution methods
Jayarathna et al. [57]	2020	41	Survey of the VRP with multiple depots
			Mathematical formulation
			Summary of 16 papers from 2011 to 2020 that use exact solution methods
			Summary of 19 papers from 2004 to 2020 that use heuristic and metaheuristic solution methods
Karakatič and Podgorelec [59]	2015	91	Genetic algorithms for the multi-depot VRP
			Strengths and weaknesses of methods, operators, and settings
			Computational comparison of genetic operators on benchmark problems
			Computational comparison of two genetic algorithms to nine solution methods (e.g., tabu search) on five benchmark problems

(continued)

Table 1.12 (continued)

Article	Year	Reference count	Comments
Montoya-Torres et al. [86]	2015	183	Survey of the VRP with multiple depots
			Review of papers from 1988 to 2014
			Mathematical model
			Detailed tabular summary of 130 papers with single objective and 17 papers with multiple objectives
			Type of variant including time windows, heterogeneous fleet, capacitated, periodic, pickup and delivery, and split delivery
			Solution method
			Exact, heuristic, and metaheuristic

Table 1.13 Pickup and Delivery and Dial-a-Ride Problems

Article	Year	Reference count	Comments
Koç et al. [62]	2021	124	VRP with simultaneous pickup and delivery
			Classification of pickup and delivery problems
			Mathematical models
			Exact algorithms
			Heuristics including construction and improvement, local search, population search, and ant colony systems
			Variants including stochastic, green, and heterogeneous problems, time windows, and multiple depots
			Case studies
			Computational comparison of metaheuristics
Ho et al. [51]	2018	115	Survey of dial-a-ride problems
			Problem history
			Application areas
			Survey of 86 papers published since 2007
			Problem characteristics (static, dynamic)
			Solution methods (exact, heuristic)
			Benchmark instances and results
			Computational comparison of recent algorithms
			New technologies and trends
			New research directions

(continued)

Table 1.13 (continued)

Article	Year	Reference count	Comments
Koç and Laporte [61]	2018	107	VRP with backhauls
			Mathematical models
			Exact algorithms
			Heuristics including local-search metaheuristics, population search, and neural networks
			Variants including time windows, multiple depots, and heterogeneous fleet
			Case studies and industrial applications
			Computational comparison of recent metaheuristics
Potvin [107]	2009	195	Evolutionary algorithms
			Genetic algorithms for the VRP with backhauls and with simultaneous pickup and delivery, and the pickup and delivery problem with time windows
			Computational results for evolutionary algorithms on the pickup and delivery problem with time windows
Parragh et al. [102]	2008	109	Survey on pickup and delivery problems
			Focus on transportation of good from the depot to the customers (linehauls) and from the customers to the depot (backhauls)
			Several variants of the VRP with backhauls including clustered backhauls (all linehauls before backhauls) and any sequence of linehauls and backhauls
			Mathematical formulations
			Solution approaches (exact, heuristic, metaheuristic)
			Benchmark instances
Parragh et al. [103]	2008	217	Survey on pickup and delivery problems
			Focus on transportation of goods between pickup and delivery locations
			VRP with pickups and deliveries (paired and unpaired pickup and delivery points, dial-a-ride problem, static, dynamic, and stochastic versions)
			Mathematical formulations
			Solution approaches (exact, heuristic, metaheuristic)
			Benchmark instances
Berbeglia et al. [11]	2007	148	Survey on pickup and delivery problems
			Classification scheme
			Many-to-many problems including the swapping problem
			One-to-many-to-one problems including the TSP with pickups and deliveries
			One-to-one problems including the VRP with pickups and deliveries and the dial-a-ride problem
			Discussion of exact algorithms and heuristic methods for each type of problem

(continued)

Table 1.13 (continued)

Article	Year	Reference count	Comments
Cordeau and Laporte [27]	2007	44	Dial-a-ride problem Main features (e.g., static, dynamic) Mathematical models Single-vehicle and multi-vehicle problems (static, dynamic) Exact and heuristic methods for each type of problem

Table 1.14 Rich and Multi-attribute Vehicle Routing

Article	Year	Reference count	Comments
Lahyani et al. [68]	2015	209	Rich vehicle routing problems: taxonomy and definition 41 selected papers
Caceres-Cruz et al. [19]	2014	123	Survey of rich VRPs Definition and history of the VRP Variants of the VRP Solution methods Definition of rich VRP Rich VRP literature review Classification of rich VRP papers Insights and trends
Vidal et al. [129]	2013	279	Survey of heuristics (and metaheuristics) for multi-attribute VRPs VRP heuristics Relative performance of heuristics Classification of heuristics Frequently encountered attributes Winning solution strategies Conclusions and perspectives
Drexl [31]	2012	86	Overview of rich vehicle routing in theory and practice Rich VRPs have many dimensions Requests Fleet Route structure Objectives Scope of planning Applications Solution methods Trends in VRP research Commercial vehicle routing software Gaps between theory and practice

Table 1.15 Routing over time

Article	Year	Reference count	Comments
Wang and Wasil [131]	2021	79	Periodic vehicle routing problems Summaries of seven articles published in Networks from 1974 to 2014
Mor and Speranza [88]	2020	94	Vehicle routing problems over time Classification of the vehicle routing problems by the decisions to be taken
			Assignment and sequencing: Capacitated vehicle routing
			Customer selection: Vehicle routing with profit
			Demand to deliver: Vehicle routing with split deliveries
			Commodity: Vehicle routing with multiple commodities
			Time
			Periodic routing problems
			Inventory routing problems
			Vehicle routing problems with release dates
			Multi-trip vehicle routing problems
Moons et al. [87]	2017	75	Production scheduling-vehicle routing problem Classification scheme based on production, inventory, and distribution characteristics
			Literature review of 33 papers published between 1996 and 2016
			Problem characteristics
			Solution approaches
			Research directions
			Real-life production, inventory, and distribution characteristics and objective criterion
			Uncertainty
			Solution algorithms
			Value of integration and sensitivity analysis

(continued)

Table 1.15 (continued)

Article	Year	Reference count	Comments
Paraskevopoulos et al. [99]	2017	104	Resource constrained routing and scheduling Basic models Skill VRP and its variants Technician routing and scheduling problem and its variants Applications Home healthcare Installation, maintenance, and repairs Forest management Airport operations Three-field taxonomy Resource qualification Service requirements Planning horizon and temporal constraints Precedence Importance Objectives Priorities Outsourcing and overtime Workload balancing Service completion time and delays Other or several objectives Solution methods Exact algorithms Heuristic algorithms Matheuristics and decomposition algorithms Stochastic programming and robust optimization algorithms Research directions

(continued)

Table 1.15 (continued)

Article	Year	Reference count	Comments
Gendreau et al. [45]	2015	98	Time dependent routing problems Applications of time dependent routing Route planning in road networks Travel planning in public transit networks Vehicle routing applications including aircraft, ship, and submarine route planning Robot motion planning Travel time and speed models Deterministic models Stochastic models Time dependent point-to-point route planning Quickest path problem Least consumption path problem Stochastic quickest path problem Time-dependent traveling salesman problem Time-dependent vehicle routing problem Static and deterministic problem Static and stochastic problem Dynamic problem Time-dependent arc routing problem Future research directions

(continued)

Table 1.15 (continued)

Article	Year	Reference count	Comments
Campbell and Wilson [20]	2014	92	Periodic vehicle routing problem Introduce the periodic vehicle routing problem Problem definition Three important early contribution in 1974, 1979, and 1984 Basic variants: the periodic traveling salesman problem, the periodic vehicle routing problem with time windows, and the multidepot vehicle routing problem Applications of the periodic vehicle routing problem Pickup Delivery On-site service Solution methods Early heuristics Metaheuristics including tabu search and variable neighborhood search Other variants including problems with intermediate facilities, location routing, reassignment constraints, and alternative objective functions Future directions Variants Multiobjective Stochasticity
Bräysy and Gendreau [17]	2005	73	Vehicle routing problems with time windows Introduce the vehicle routing problem with time windows Evaluation criteria for heuristics: computational time, solution quality, ease of implementation, robustness, and flexibility Construction heuristics Computational comparison of three selected heuristics from 1987 to 2001 Improvement Methods Literature review Computational comparison of 11 selected heuristics from 1993 to 2003

(continued)

Table 1.15 (continued)

Article	Year	Reference count	Comments

Bräysy and Gendreau [18]	2005	102	Vehicle routing problems with time windows
			Tabu search
			    Table summarizing the main features of 14 tabu search algorithms from 1994 to 2003 (initial solution, neighborhood operators, whether route number reduction is used)
			    Computational comparison of 12 tabu search algorithms
			Genetic algorithms
			    Table summarizing the main features of 17 genetic algorithms and evolution strategies from 1993 to 2005 (initial population, crossover, mutation)
			    Computational comparison of 14 algorithms
			Other metaheuristics such as greedy randomized adaptive search, simulated annealing, ant colony optimization, and variable neighborhood descent
			    Computational comparison of 19 algorithms from 1994 to 2004
			Discussion
			    Top 10 metaheuristics
			    Trade-off between computational time and cumulative number of vehicles used

Table 1.16 Shipping

Article	Year	Reference count	Comments
Zis et al. [135]	2020	144	Taxonomy and survey of ship weather routing
			Most articles have appeared in past 20 years
			Optimization criterion
			Shipping sector
			Geography
			Solution approach
			Weather data
			Resolution of data
			Ship fuel efficiency modeling
			Emissions considered
			Size of fleet
Meng et al. [83]	2014	94	Overview of containership routing and scheduling in liner shipping
			Reviews the past 30 years
			Strategic problems in liner ship routing
			Tactical decisions
			Operational problems
			Future (practical) research perspectives
Psaraftis and Kontovas [109]	2013	72	Survey of speed models in maritime transportation
			Who is the speed optimizer?
			What is being optimized?
			Non-emissions speed models
			Emissions speed models
			Does imposing speed limits make sense?

Table 1.17 Two-echelon, Collaborative, and Inter-terminal Problems

Article	Year	Reference count	Comments
Mavi et al. [82]	2020	72	Systematic literature review of cross-docking Definition, advantages, and applications Techniques used for cross-docking optimization Vehicle routing/inventory management is a key component
Gansterer and Hartl [44]	2018	86	Survey on collaborative vehicle routing Centralized collaborative planning Decentralized planning without auctions Decentralized planning with auctions Future research directions
Heilig and Voss [50]	2017	91	Annotated bibliography on inter-terminal transportation Vehicle routing plays a key role in coordinating inter-terminal flows This connection has been under-studied Green logistics is applicable here
Cuda et al. [28]	2015	57	Survey on two-echelon routing problems Two-echelon distribution systems Two-echelon location routing problems Two-echelon vehicle routing problems Truck and trailer routing problems Open research areas
Drexl [32]	2012	120	Survey of synchronization in vehicle routing What does synchronization mean? Concrete example Classification scheme Modeling and algorithmic issues Literature survey Related fields

Table 1.18 Specific Variants, Benchmark Datasets, and Software

Article	Year	Reference count	Comments
Gutiérrez-Sánchez and Rocha-Medina [48]	2022	0	VRP variants applicable to donation collection problems
			Review and taxonomy of applicable VRP models
			VRP with time windows, VRP with backhauls, VRP with pickups and deliveries, and periodic VRP
			Classification of solution approaches
			Exact methods
			Heuristic methods
			Metaheuristic methods
Cheikhrouhou and Khoufi [23]	2021	34	Survey of the multiple traveling salesman problem (MTSP)
			Applications of the MTSP
			Transportation and delivery
			Monitoring and surveillance
			Cooperative mission
			Disaster management
			Precision agriculture
			Search and rescue
			Wireless sensor network data collection network and connectivity
			Multi-robot task allocation and scheduling
			MTSP variants by
			Salesmen characteristics
			Depot specifications
			Cities specifications
			Objective function
			Problem constraints
			Solution approaches
			Exact methods
			Metaheuristics
			Market-based methods including central and distributed auctioneer
			Other methods such as game theory and fuzzy logic
			Taxonomy based on MTSP variants, solution approaches, and application domains
			Future directions
Gunawan et al. [47]	2021	139	Review of vehicle routing benchmark datasets
			Analysis of VRP publications
			Identification of 41 VRP variants
			Summary of 131 datasets from 1959 to 2020
			Brief descriptions of datasets for VRP variants including time windows, green, and dynamic problem with time windows

(continued)

Table 1.18 (continued)

Article	Year	Reference count	Comments
Cattaruzza et al. [21]	2018	95	VRPs with multiple trips
			Mathematical formulations of the multi-trip VRP
			4-index
			3-index with vehicle index and without trip index
			3-index with trip index and without vehicle index
			2-index without vehicle index and trip index
			Benchmark instances
			Exact and heuristic solution approaches
			Problem variants
			Time windows
			Service-dependent loading time
			Limited trip duration
			Profits (serving all customers is not mandatory)
			Formulations
			Benchmark instances
			Exact and heuristic solution approaches
			Production-routing, inventory-routing, production-inventory-routing problems
			Multi-trip in maritime transportation
			Multi-level distribution with multi-trips
			Other variants
			Original temporal constraints
			Distribution of incompatible products
			Limited number of trips
			Fleet size minimization
			State-of-the-art results on benchmark instances
Rincon-Garcia et al. [112]	2018	102	Vehicle routing software
			Evolution of computerized vehicle routing and scheduling software from the 1970s to the 2000s
			Software characteristics of five main vendors in the United Kingdom including routing functions, algorithm considerations, and type of fleet
			Survey of logistics providers in the United Kingdom
			Nine main reasons to adopt software
			Impact of new challenges in the industry
			Evaluation of software capabilities
			Literature review
			Data concerning traffic information
			Discussion of the VRP, algorithms, heuristics, and 10 variants (e.g., time windows)
			Time-dependent VRP
			Discussion of academic theory, commercial VRP, and industry needs

(continued)

Table 1.18 (continued)

Article	Year	Reference count	Comments
Koç et al. [60]	2016	140	Survey of heterogeneous vehicle routing
			Classification of heterogeneous VRPs
			Problem definition
			Mathematical formulations (single-commodity flow, two-commodity flow, set partitioning)
			Fleet size and mix VRP without and with time windows
			Heuristics (population search, tabu search)
			Heterogeneous fixed fleet VRP
			Tabu search and other heuristics
			Classification of more than 30 heterogeneous VRP variants including multiple depots, backhauls, green, split deliveries, and pickup and delivery
			Eight real-life case studies
			Detailed tabular summary of the literature on the heterogeneous VRP
			Detailed comparative analysis of the computational performance of state-of-the-art metaheuristics
			Future research directions
Kovacs et al. [64]	2014	118	Survey of service consistency in vehicle routing
			Application areas (consistency in small package shipping, consistency apart from customer satisfaction)
			Problem description
			Early approaches to deal with fluctuating demand
			A priori routing
			Districting
			Demand stabilization
			Modeling and solving problems with arrival time consistency; with person-oriented consistency
			Inconsistency
Archetti and Speranza [5]	2012	63	Survey of VRP with split deliveries
			Problem description and formulation
			Known problem properties
			Computational complexity
			Reducibility
			k-split cycles
			Number of splits
			Savings
			Heuristic algorithms
			Exact algorithms
			Variants including time windows, pickup and delivery, profit maximization, inventory and production, minimum fraction served, heterogeneous fleet, stochastic demands, discrete demands, and arc routing
			Applications

(continued)

Table 1.18 (continued)

Article	Year	Reference count	Comments
Vansteenwegen et al. [128]	2011	64	Survey of the orienteering problem
			Orienteering problem and orienteering problem with time windows
			Definition of both problems and mathematical formulations
			Practical applications including home fuel delivery
			Benchmark instances
			Solution approaches
			Team orienteering problem and team orienteering problem with time windows
			Definition of both problems and mathematical formulations
			Practical applications
			Benchmark instances
			Solution approaches
			Variants of the orienteering problem
			Future research lines including arc routing version and dealing with capacitated vehicles

Table 1.19 Exact Algorithms and Heuristics

Article	Year	Reference count	Comments
Wang and Wasil [131]	2021	79	Exact methods
			Descriptions of 13 articles published in Networks from 1981 to 2011
			Branch-and-bound
			Branch-and-cut
			Column generation
			Heuristics
			Descriptions of three articles published in Networks from 1977 to 1993
			VRP implementation issues
			Modified Clarke and Wright
			Generalized assignment
			Tabu search

(continued)

Table 1.19 (continued)

Article	Year	Reference count	Comments
Elshaer and Awad [36]	2020	325	Metaheuristics
			Eight single solution-based methods including simulated annealing, tabu search, and variable neighborhood search
			Sixteen population-based methods with 10 evolutionary computation methods and six swarm intelligence methods
Ansari et al. [3]	2018	151	Continuous approximation models
			Freight distribution systems with and without transshipment including one-to-many, many-to-one, and many-to-many classes
Masutti and de Castro [80]	2017	9	Bee-inspired algorithms
			Bee colonies as swarm intelligence
			Standard bee-inspired algorithm approaches
			Artificial bee colony optimization (ABO)
			Bee colony optimization (BCO)
			Bee systems (BS)
			Marriage in honeybees optimization (MHBO)
			Literature review on the standard bee-inspired algorithms
			33 articles from 2010 to 2016 on ABO
			Nine articles from 2008 to 2012 on BCO
			Six articles from 2002 to 2015 on BS
			Six articles from 2007 to 2013 on MHBO
			TSPoptBees algorithm for solving the TSP
Kritzinger et al. [65]	2015	94	Adaptive search mechanisms with metaheuristics
			Adaptive tabu search, guided and iterated local search, adaptive variable neighborhood and large neighborhood search
			Population-based methods including ant colony optimization and genetic algorithms

(continued)

Table 1.19 (continued)

Article	Year	Reference count	Comments
Kritzinger et al. [66]	2015	22	Adaptive search mechanisms with metaheuristics
			Tested six adaptive mechanisms using two variants of adaptive variable neighborhood search on instances of the open VRP with and without time windows
Archetti and Speranza [6]	2014	92	Matheuristics for routing problems
			Decomposition approaches
			Improvement heuristics
			Branch-and-price/column generation-based approaches
Baldacci et al. [9]	2012	29	Exact algorithms
			VRP with capacity constraints (branch-and-cut algorithms, algorithms based on set partitioning formulation)
			VRP with time windows
			Computational comparison of exact methods for both problems
Potvin [107]	2009	195	Evolutionary algorithms
			Genetic algorithms
			Evolution strategies
			Particle swarm optimization
			Classification by different problem types
			Computational performance on the VRP, VRP with time windows, and pickup and delivery problem with time windows
Baldacci et al. [10]	2007	41	Exact algorithms for the capacitated VRP
			Mathematical formulations
			Relaxations and valid inequalities
			Branch-and-cut methods
			Report on the computational performance of three effective exact methods
Funke et al. [43]	2005	66	Local search for vehicle routing and scheduling problems
			Generic local search
			Neighborhood types
			Search techniques
			Optimization-based indirect local search

Table 1.20 Applications

Article	Year	Reference count	Comments
Euchi et al. [38]	2022	101	Review of home health care routing and scheduling problems
			67 relevant papers are summarized
			Taxonomy includes
			Characteristics of papers
			Objective functions
			Solution methods
			Benchmark instances
			Constraints
			Future work
			Integration with new technologies (e.g., drones)
Fleckenstein et al. [41]	2022	123	Integrating demand management and vehicle routing
			Generalized problem definition
			Request capture
			Demand management
			Order confirmation
			Vehicle routing
			Mathematical model
			Dynamic control model that incorporates vehicle routing decisions
			Solution concepts
			Decomposition-based approximation that can check route feasibility with heuristic and exact algorithms and develop a full route plan
			Detailed overview of 55 articles from 2005 to 2022 that covers 12 areas including type of application and type of routing problem (e.g., deliveries, pickups, time windows); for 43 articles, a detailed overview of the decomposition-based approach used including the type of feasibility check (e.g., route-based, capacity-based)
			Discussion of future research topics including generic model formulations, advancing solution approaches, and transferring existing methods into practice

(continued)

Table 1.20 (continued)

Article	Year	Reference count	Comments
Fröhlich et al. [42]	2022	114	Safe and secure vehicle routing
			Review articles in five categories with more than 50% appearing in 2016 to 2021
			Transportation of hazardous materials
			12 articles that span 10 problem types such as VRP with time windows and are solved with 10 exact and heuristic methods including tabu search
			Patrol routing (e.g., guards and police officers)
			19 articles that span 12 problem types such as team orienteering with time windows and are solved with 12 exact methods and heuristics including genetic algorithms
			Cash-in-transit (e.g., money is picked up from or delivered to banks)
			21 articles that span 13 problem types such as the periodic vehicle routing problem with time windows and are solved with 18 exact and heuristic methods including branch-and-cut
			Dissimilar routing problems
			28 articles that cover three problem types such as the m-peripatetic TSP and VRP and are solved with 10 exact methods and heuristics including tabu search
			Modeling of multi-graphs
			15 articles that develop models when multiple attributes (e.g., cost and travel time) are defined on the arcs
			Summary of datasets (e.g., real-world and artificial instances) used in routing hazardous material, patrol routing, and cash-in-transit work
			Extensive discussion of future research directions and possible extensions

(continued)

Table 1.20 (continued)

Article	Year	Reference count	Comments
Utamima and Djunaidy [127]	2022	46	Agricultural routing planning
			Coverage planning problem (CPP) (VRP in farm logistics)
			Aerial and mini robots
			Crop harvesting
			Biomass collection
			Field-track generation
			Agricultural routing planning (ARP)
			Minimize non-working distance of machines in the field
			Two variations including consideration of obstacles in the field
			Minimize the time to complete field operations
			Three variations including minimizing the time of the last vehicle to finish its field operations
			Applications of ARP
			Weed control operations
			Fertilizing
			Cultivation, mowing
			Orchard operations
			Wildlife avoidance
			Mathematical model
			Heuristic methods
			Clarke and Wright, tabu search, genetic algorithm
Zennaro et al. [133]	2022	344	Identify vehicle routing problems that are related to e-commerce in outbound and reverse logistics
			Present 25 articles that deal with 14 delivery and routing problems including the dynamic VRP, the VRP with multiple trips and time windows, and the pickup and delivery problem with time windows
Anuar et al. [4]	2021	84	Vehicle routing in humanitarian operations
			Total of 123 papers from 2010 to 2020 are reviewed
			Descriptive analysis (charts and tables) of papers (e.g., number of papers by year of publication, type of publication, source of publication)
			Comprehensive review of the literature: Supply and delivery in routing optimization problems, evacuation VRPs, rescue in routing optimization problems
			Machine learning, exact, heuristic, local search, metaheuristic, and hybrid methods
			Extensive analysis of current trends (e.g., 73 papers focus on supply and delivery problems) presented in 36 figures (e.g., pie charts)
			Five questions for future research (e.g., Can a general model be developed that applies to all disaster phases and types?)

(continued)

Table 1.20 (continued)

Article	Year	Reference count	Comments
Liang et al. [71]	2021	63	Waste collection routing problem
			Review of 35 articles published from January 2014 to January 2020
			Summary of heuristics used to solve real-world waste collection problems in 13 articles
			Summary of heuristics based on problem characteristics (objective function, constraints, and vehicle type) and 13 heuristics including ant colony optimization, genetic algorithm, and tabu search
			Summary of heuristics on types of problems solved including waste collection routing, split deliveries, routing with stochastic demand, and periodic arc routing
			Comparison of results on 10 benchmark instances reported in five articles
			Mathematical formulations of the waste collection routing problem with time windows and lunch break, split-delivery VRP, asymmetric capacitated VRP, and capacitated VRP with stochastic demand
			Discussion of the waste collection arc routing problem and periodic capacitated arc routing problem with mathematical formulations
			Geographic information system for the waste collection routing problem
			Combined with a heuristic
			Applications in waste collection
Odum et al. [93]	2021	26	Use of disruption management in vehicle routing
			Article focuses on the problem in which a vehicle breaks down and a new solution needs to be generated quickly
			Review articles in nine areas
			Ant colony optimization
			Heuristics for delivery problems
			Waste collection
			Arc routing
			Node routing
			Container delivery and collection
			Construction and improvement algorithms for the VRP
			Metaheuristics
			Simulated annealing
			Tabu search
			Genetic algorithm
			Cluster analysis
			Hard and soft clusters
			K-means
			Probabilistic distance clusters

(continued)

Table 1.20 (continued)

Article	Year	Reference count	Comments
Peker and Türsel Eliiyi [104]	2021	82	Shuttle bus service routing Review of 75 articles published from 2000 to 2020 Classification by Journal outlets Application areas including employee/personnel, patient/hospital, students, elderly/disabled, and airport shuttles Objective functions General approach and specific solution methods
Ellegood et al. [35]	2020	109	School bus routing problem Classification of the school bus routing problem literature Sub-problem type Number of schools Service environment Load type Fleet mix Objectives Constraints Solution approaches Classical heuristics, construction heuristics, improvement heuristics, exact methods, metaheuristics including population-based evolutionary approaches and trajectory-based methods, and methods addressing uncertainty for the bus route generation sub-problem
Utama et al. [126]	2020	84	VRP for perishable goods Review of 35 articles published from January 2014 to January 2020 Discussion and detailed tables summarizing 42 papers with a single objective and 17 papers with multiple objectives and type of solution approach used (heuristic, metaheuristic, exact, hybrid, simulation) Detailed charts that describe and summarize the distribution of articles (e.g., by type of objective (single, multiple)) over the years
Fikar and Hirsch [40]	2017	67	Home health care routing and scheduling Review of 25 articles focusing on single-period problems and 19 articles focusing on multi-period problems including solution method used (e.g., metaheuristic), objectives (e.g., travel time, travel distance, workload balance), and constraints (e.g., time windows) Discussion of four future research directions Stochastic routing and scheduling Integrated multi-stage, multi-period planning approaches Multimodality and mode of transport choices Sustainability considerations and acceptance of optimization methods

(continued)

Table 1.20 (continued)

Article	Year	Reference count	Comments
Coelho et al. [25]	2016	91	Survey of real VRP applications
			Five key areas from 2000 to 2015
			Detailed discussion and summary table of articles in each area that list year, journal, algorithm, product, and estimated improvement
			Areas of application
			Oil, gas, and fuel (eight articles)
			Retail (eight articles)
			Waste collection and management (15 articles)
			Mail and small package delivery (six articles)
			Food distribution (23 articles)
			Identified three main research trends
			Green transportation
			Routing with time-dependent or congestion-related travel time
			Quality of service as an objective or constraint
Han and Ponce-Cueto [49]	2015	97	Waste collection VRP
			Discussed three categories of waste collection problems
			Household (five arc routing papers from 1995 to 2012)
			Commercial (six papers from 1996 to 2012)
			Roll-on-roll-off (five papers from 1997 to 2013)
			Discussed the main contributions and methods to solve the waste collection VRP in 33 node routing papers from 1996 to 2014; in 16 arc routing papers from 1974 to 2012; in 16 papers with both types of models from 1989 to 2014
Park and Kim [100]	2010	72	School bus routing problem
			Detailed discussion of five sub-problems (data preparation, bus stop selection, bus route generation, school bell time adjustment, route scheduling) in 29 papers from 1969 to 2009
			Classification of 26 papers based on eight problem characteristics including number of schools (single or multiple), fleet mix, objectives (e.g., maximum route length), and constraints (e.g., maximum riding time)
			Classification of 15 papers based on the mathematical model (e.g., mixed integer program)
			Discussion of metaheuristics that have been applied to the school bus routing problem
			Future research directions

1.3   Observations and Trends

To help identify trends in VRP research and practice, we tallied the number of recent survey articles over the past five years or so (2017 to June 2022) by category. These counts are given in Table 1.21. We tallied the number of citations for each of the 135 articles from the date of publication until June 2022. The survey articles with 500 or more citations are given in Table 1.22.

Over the past five years, the most survey articles are in applications with 12 articles and green routing (including electric vehicle routing) with 12 articles, about 31% of the

Table 1.21 Number of VRP survey articles published from 2017 to 2022 by category

Category	Number of articles	Percentage
Applications	12	15.3
Green Routing	12	15.3
Drones, Last-mile Delivery, and Urban Distribution	8	10.3
Overview	7	9.0
Dynamic and Stochastic Routing	6	7.7
Specific Variants, Benchmark Datasets, and Software	5	6.4
Exact Algorithms and Heuristics	4	5.1
Routing Over Time	4	5.1
Arc Routing and General Routing	3	3.8
Pickup and Delivery and Dial-a-Ride Problems	3	3.8
Two-echelon, Collaborative, and Inter-terminal Problems	3	3.8
Alternative and Multiple Objectives	2	2.6
Inventory Routing	2	2.6
Loading Constraints	2	2.6
Location-Routing	2	2.6
Multiple Depots	2	2.6
Shipping	1	1.3
Rich and Multi-attribute Routing	0	0

Table 1.22 Highly cited VRP survey articles

Number of Citations	Authors	Topic	Year	Journal
1550	Bräysy and Gendreau [17]	Time windows: Route construction and local search algorithms	2005	Transportation Science
1338	Pillac et al. [105]	Dynamic VRP	2013	European Journal of Operational Research
1202	Nagy and Salhi [91]	Location-routing	2007	European Journal of Operational Research
1157	Laporte [70]	Fifty years of vehicle routing	2009	Transportation Science
1087	Bräysy and Gendreau [18]	Time windows: Metaheuristics	2005	Transportation Science
1031	Vansteenwegen et al. [128]	Orienteering problem	2011	European Journal of Operational Research
983	Cordeau and Laporte [27]	Dial-a-ride problem	2007	Annals of Operations Research
962	Braekers et al. [16]	State-of-the-art classification and review of the VRP	2016	Computers & Industrial Engineering
962	Eksioglu et al. [34]	Taxonomic review of the VRP	2009	Computers & Industrial Engineering
899	Lin et al. [73]	Green vehicle routing problem	2014	Expert Systems with Applications
889	Berbeglia et al. [11]	Static pickup and delivery problems	2007	TOP
704	Demir et al. [30]	Recent research on green road freight transportation	2014	European Journal of Operational Research
675	Prodhon and Prins [108]	Location-routing	2014	European Journal of Operational Research
675	Coelho et al. [24]	Thirty years of inventory routing	2014	Transportation Science
610	Andersson et al. [2]	Combined inventory management and routing	2010	Computers & Operations Research
591	Jozefowiez et al. [58]	Multi-objective VRPs	2008	European Journal of Operational Research

78 published articles. Twenty-one of the 24 articles were published in 2020, 2021, and 2022. The next most popular areas are drones, last-mile delivery, and urban distribution, overviews, and dynamic and stochastic routing with 8, 7, and 6 articles, respectively. Rich and multi-attribute vehicle routing has no survey article published since 2015.

Citation counts provide an indication of the continuing interest in an area. From Table 1.22, we see that the two-part series on time windows by Bräysy and Gendreau [17, 18] is very highly cited with a total of 2,637 citations. The remaining four articles with more

Table 1.23 Highly recommended VRP survey articles

Authors	Year	Area	Comments
Boysen et al. [15]	2021	Last-mile delivery	Forward looking
Gunawan et al. [47]	2021	Benchmark datasets	Useful, unique
Macrina et al. [76]	2020	Drones	Important topic
Vidal et al. [130]	2020	Emerging variants	Forward looking
Rossit et al. [116]	2019	Visual attractiveness	Unique
Ansari et al. [3]	2018	Continuous approximation models	Unique
Gansterer and Hartl [44]	2018	Collaborative routing	Unique
Matl et al. [81]	2018	Workload equity	Unique
Otto et al. [96]	2018	Drones	Important topic
Mourão and Pinto [89]	2017	Arc routing	Comprehensive
Psaraftis et al. [110]	2016	Dynamic routing	Important topic
Campbell and Wilson [20]	2014	Periodic routing	Nicely done
Kovacs et al. [64]	2014	Consistency	Unique
Archetti and Speranza [5]	2012	Split deliveries	Unique
Bertazzi and Speranza [13]	2012	Inventory routing	Nice introduction
Drexl [32]	2012	Synchronization in routing	Unique
Laporte [70]	2009	Fifty years of routing	Good general survey

than 1,000 citations are in dynamic routing [105], location-routing [91], a 50-year overview of the VRP [70], and the orienteering problem [128].

We would like to recommend the 17 articles given in Table 1.23. We found these articles to be unique, forward looking, or important. The coverage in the most recent articles since 2020 (last-mile delivery, drones, emerging variants) provides a strong indication of the trending topics currently being pursued by researchers and practitioners.

Finally, the survey articles covered in this book mention dozens of future research topics. We list some of these below. We are hopeful that readers (in particular Ph.D. students) will find our list helpful.

1. Many real-world vehicle routing problems are dynamic and stochastic. Research papers should reflect this.

2. More than ever before, historical data are available. Researchers can use machine learning and pattern recognition to take advantage of these data.

3. Vehicle routing optimization has recently been applied to humanitarian operations. Most published research involves delivering critical aid to victims of disasters. More attention should be paid to evacuation and rescue.

4. How can trucks, drones, and robots be used, in combination, to improve delivery operations? What strategies work best in urban, suburban, and rural settings?

5. In last mile delivery, a large territory must often be partitioned into smaller regions such that each region has its own dedicated vehicles. How does territory design change when trucks, drones, and robots can make deliveries, in combination?

6. How should vehicle routing models change based on the level of real-time transportation information?

7. In the area of green vehicle routing, charging vehicles in peak hours can be time consuming and lead to significant queueing issues. What role, if any, can scheduling play in reducing long waiting times?

8. In order to handle a wide variety of real-world costs and constraints, as well as uncertainty, simheuristics (simulation plus metaheuristics) might play a more prominent role in vehicle routing research. What would some examples of this look like?

9. With respect to robot-driven deliveries, one key question is: How should these services be priced? What is the objective? Is it to maximize profit or increase customer demand and acceptance of the new technology? Does dynamic pricing make sense here and, if so, how does this interact with choice of delivery time windows?

10. When trucks, drones, and robots are used, in combination, to make deliveries, how do we determine an optimal or near-optimal fleet size and mix? Assume that the objective is to minimize overall delivery time and costs. Alternatively, now suppose a secondary objective is to minimize greenhouse gas emissions. How would the ideal fleet size and mix change?

11. Arc routing problems under uncertainty may involve many different types of uncertainty. For example, if we focus on risk, we might consider the use of drones for surveillance and monitoring in high risk areas. How do we measure the risk of a route? How do we search for and obtain optimal or near-optimal routes?

12. Vehicle routing with backhauls has been studied since the 1980s. Most of the solution procedures have been heuristics. Can effective exact algorithms be developed? To our knowledge, a stochastic version of this problem has not yet been considered.

13. In studying the electric vehicle routing problem, the total cost should include driver costs, vehicle costs, and energy costs. What should an appropriate and realistic objective function look like in this setting? If the objective function or some constraints become nonlinear, do linear approximations make sense? Can we solve these problems in reasonable running times?

14. The truck and drone routing problem has been well-studied in the static case, where demands are known in advance. In an on-demand delivery system, requests for ser-

vice arrive dynamically over the course of a planning horizon. This topic merits more research attention.

15. Even in the static truck and drone routing problem, improved exact algorithms are needed to solve instances of practical size.

16. In recent years, several authors have proposed a variety of measures of the visual attractiveness of vehicle routes. In future work, researchers might try to develop and evaluate new objective functions that combine traditional measures such as route length and the number and cost of vehicles with successful measures of visual attractiveness.

17. We assume that the majority of the readers of this book are academics with a background in operations research. A key measure of our success as vehicle routing researchers is the extent to which our work has a positive, practical impact.

It is well-known that vehicle routing software products have been used by small, medium, and large companies in North America and Europe since the 1980s. Many of these products are powerful, versatile, and incredibly helpful. They have been developed mainly by talented computer scientists and operations researchers, but very few of these products use Gurobi or CPLEX. In the next few years, there is a wonderful opportunity to develop tools and solution methods that do not require a deep knowledge of integer programming and yet enable software developers to improve and enhance the ability of commercial vehicle routing software to respond to a variety of emerging and more demanding requests from industry. It is essential that vehicle routing research continues to be impactful.

18. There has been a small amount of research on parallel algorithms for the capacitated vehicle routing problem. This research can and should be extended to more complicated vehicle routing variants.

19. Since 2013, 10 survey articles on dynamic and stochastic vehicle routing have been published (see Table 1.7). Four have been published since 2020. There are many opportunities for future work on this topic. For example, there are numerous ways to model the reliability and availability of information over time.

20. The large number and breadth of recent VRP applications are impressive with new contributions in diverse areas such as home health care, agriculture, and safe and secure routing. We expect the application areas will continue to grow and can only imagine what new areas will be explored (e.g., routing of passenger drones or flying taxis for daily commuting).

1.4   Conclusions

We uncovered 135 survey articles published since 2005 that span a wide range of areas in vehicle routing. We categorized these articles and summarized their contributions in 18 detailed tables. We identified trending areas of research and practice over the past five years including green routing, drones, last-mile delivery, and urban distribution, and dynamic and stochastic routing. We recommended 17 articles that we found to be unique, forward looking, or important. We hope that our compilation provides valuable information to those just starting down the road of vehicle routing research or practice or to those seeking a new road to travel.

Finally, although we tried to be exhaustive in our search, we know that there are survey articles that we missed, and we apologize for not including them here. However, we continue to amass new articles, so please do send them our way.

Acknowledgements We thank Janet Cavanagh at the University of Maryland for her skillful help in finalizing this book.

References

1. L. ALFASEEH AND B. FAROOQ, *Multi-factor taxonomy of eco-routing models and future outlook*, Journal of Sensors, 2020 (2020). Article 4362493.
2. H. ANDERSSON, A. HOFF, M. CHRISTIANSEN, G. HASLE, AND A. LOKKETANGEN, *Industrial aspects and literature survey: Combined inventory management and routing*, Computers & Operations Research, 37 (2010), pp. 1515–1536.
3. S. ANSARI, M. BAŞDERE, X. LI, Y. OUYANG, AND K. SMILOWITZ, *Advancements in continuous approximation models for logistics and transportation systems: 1996–2016*, Transportation Research Part B, 107 (2018), pp. 229–252.
4. W. K. ANUAR, L. S. LEE, S. PICKL, AND H.-V. SEOW, *Vehicle routing optimisation in humanitarian operations: A survey on modelling and optimisation approaches*, Applied Sciences, 11 (2021). Article 667.
5. C. ARCHETTI AND M. G. SPERANZA, *Vehicle routing problems with split deliveries*, International Transactions in Operational Research, 19 (2012), pp. 3–22.
6. C. ARCHETTI AND M. G. SPERANZA, *A survey on matheuristics for routing problems*, EURO Journal on Computational Optimization, 2 (2014), pp. 223–246.
7. M. ASGHARI AND S. M. J. M. AL-E-HASHEM, *Green vehicle routing problem: A state-of-the-art review*, International Journal of Production Economics, 231 (2021). Article 107899.
8. R. BALDACCI, E. BARTOLINI, AND G. LAPORTE, *Some applications of the generalized vehicle routing problem*, Journal of the Operational Research Society, 61 (2010), pp. 1072–1077.
9. R. BALDACCI, A. MINGOZZI, AND R. ROBERTI, *Recent exact algorithms for solving the vehicle routing problem under capacity and time window constraints*, European Journal of Operational Research, 218 (2012), pp. 1–6.
10. R. BALDACCI, P. TOTH, AND D. VIGO, *Recent advances in vehicle routing exact algorithms*, 4OR, 5 (2007), pp. 269–298.
11. G. BERBEGLIA, J.-F. CORDEAU, I. GRIBKOVSKAIA, AND G. LAPORTE, *Static pickup and delivery problems: A classification scheme and survey*, TOP, 15 (2007), pp. 1–31.
12. E. BERHAN, B. BESHAH, D. KITAW, AND A. ABRAHAM, *Stochastic vehicle routing problem: A literature survey*, Journal of Information & Knowledge Management, 13 (2014). Article 1450022.
13. L. BERTAZZI AND M. G. SPERANZA, *Inventory routing problems: An introduction*, European Journal on Transportation and Logistics, 1 (2012), pp. 301–326.

14. L. BERTAZZI AND M. G. SPERANZA, *Inventory routing problems with multiple customers*, EURO Journal on Transportation and Logistics, 2 (2013), pp. 255–275.

15. N. BOYSEN, S. FEDTKE, AND S. SCHWERDFEGER, *Last-mile delivery concepts: A survey from an operational research perspective*, OR Spectrum, 43 (2021), pp. 1–58.

16. K. BRAEKERS, K. RAMAEKERS, AND I. VAN NIEUWENHUYSE, *The vehicle routing problem: State of the art classification and review*, Computers & Industrial Engineering, 99 (2016), pp. 300–313.

17. O. BRÄYSY AND M. GENDREAU, *Vehicle routing problem with time windows, part i: Route construction and local search algorithms*, Transportation Science, 39 (2005), pp. 104–118.

18. O. BRÄYSY AND M. GENDREAU, *Vehicle routing problem with time windows, part ii: Meta-heuristics*, Transportation Science, 39 (2005), pp. 119–139.

19. J. CACERES- CRUZ, P. ARIAS, D. GUIMARANS, D. RIERA, AND A. A. JUAN, *Rich vehicle routing problem: Survey*, ACM Computing Surveys, 47(2) (2015). Article 32.

20. A. M. CAMPBELL AND J. H. WILSON, *Forty years of periodic vehicle routing*, Networks, 63 (2014), pp. 2–15.

21. D. CATTARUZZA, N. ABSI, AND D. FEILLET, *Vehicle routing problems with multiple trips*, Annals of Operations Research, 271 (2018), pp. 127–159.

22. D. CATTARUZZA, N. ABSI, D. FEILLET, AND J. GONZÁLEZ- FELIU, *Vehicle routing problems for city logistics*, EURO Journal on Transportation and Logistics, 6 (2017), pp. 51–79.

23. O. CHEIKHROUHOU AND I. KHOUFI, *A comprehensive survey on the multiple traveling sales-man problem: Applications, approaches and taxonomy*, Computer Science Review, 40 (2021). Article 100369.

24. L. C. COELHO, J.- F. CORDEAU, AND G. LAPORTE, *Thirty years of inventory routing*, Transportation Science, 48 (2014), pp. 1–19.

25. L. C. COELHO, J. RENAUD, AND G. LAPORTE, *Road-based goods transportation: A sur-vey of real-world logistics applications from 2000 to 2015*, INFOR: Information Systems and Operational Research, 54 (2016), pp. 79–96.

26. A. CORBERÁN AND C. PRINS, *Recent results on arc routing problems: An annotated bibliog-raphy*, Networks, 56 (2010), pp. 50–69.

27. J.- F. CORDEAU AND G. LAPORTE, *The dial-a-ride problem: Models and algorithms*, Annals of Operations Research, 153 (2007), pp. 29–46.

28. R. CUDA, G. GUASTAROBA, AND M. G. SPERANZA, *A survey on two-echelon routing problems*, Computers & Operations Research, 55 (2015), pp. 185–199.

29. A. DE MAIO, D. LAGANÀ, R. MUSMANNO, AND F. VOCATURO, *Arc routing under uncer-tainty: Introduction and literature review*, Computers & Operations Research, 135 (2021). Article 105442.

30. E. DEMIR, T. BEKTAŞ, AND G. LAPORTE, *A review of recent research on green road freight transportation*, European Journal of Operational Research, 237 (2014), pp. 775–793.

31. M. DREXL, *Rich vehicle routing in theory and practice*, Logistic Research, 5 (2012), pp. 47–63.

32. M. DREXL, *Synchronization in vehicle routing—a survey of vrps with multiple synchronization constraints*, Transportation Science, 46 (2012), pp. 297–316.

33. H. DÜNDAR, M. ÖMÜRGÖNÜLŞEN, AND M. SOYSAL, *A review on sustainable urban vehicle routing*, Journal of Cleaner Production, 285 (2021). Article 125444.

34. B. EKSIOGLU, A. V. VURAL, AND A. REISMAN, *The vehicle routing problem: A taxonomic review*, Computers & Industrial Engineering, 57 (2009), pp. 1472–1483.

35. W. A. ELLEGOOD, S. SOLOMON, J. NORTH, AND J. F. CAMPBELL, *School bus routing problem: Contemporary trends and research directions*, Omega, 95 (2020). Article 102056.

36. R. ELSHAER AND H. AWAD, *A taxonomic review of metaheuristic algorithms for solving the vehicle routing problem and its variants*, Computers & Industrial Engineering, 140 (2020). Article 106242.

37. T. ERDELIĆ AND T. CARIĆ, *A survey on the electric vehicle routing problem: Variants and solution approaches*, Journal of Advanced Transportation, 2019 (2019). Article 5075671.

38. J. EUCHI, M. MASMOUDI, AND P. SIARRY, *Home health care routing and scheduling problems: A literature review*, 4OR, 20 (2022), pp. 351–389.

39. J. C. FERREIRA, M. T. A. STEINER, AND O. CANCIGLIERI, *Multi-objective optimization for the green vehicle routing problem: A systematic literature review and future directions*, Cogent Engineering, 7 (2020). Article 1807082.

40. C. FIKAR AND P. HIRSCH, *Home health care routing and scheduling: A review*, Computers & Operations Research, 77 (2017), pp. 86–95.

41. D. FLECKENSTEIN, R. KLEIN, AND C. STEINHARDT, *Recent advances in integrating demand management and vehicle routing: A methodological review*, European Journal of Operational Research, (2022), https://doi.org/10.1016/j.ejor.2022.04.032.

42. G. E. A. FRÖHLICH, M. GANSTERER, AND K. F. DOERNER, *Safe and secure vehicle routing: A survey on minimization of risk exposure*, International Transactions in Operational Research, (2022), https://doi.org/10.1111/itor.13130.

43. B. FUNKE, T. GRÜNERT, AND S. IRNICH, *Local search for vehicle routing and scheduling problems: Review and conceptual integration*, Journal of Heuristics, 11 (2005), pp. 267–306.

44. M. GANSTERER AND R. F. HARTL, *Collaborative vehicle routing: A survey*, European Journal of Operational Research, 268 (2018), pp. 1–12.

45. M. GENDREAU, G. GHIANI, AND E. GUERRIERO, *Time-dependent routing problems: A review*, Computers & Operations Research, 64 (2015), pp. 189–197.

46. R. GOEL AND R. MAINI, *Vehicle routing problem and its solution methodologies: A survey*, International Journal of Logistics Systems and Management, 28 (2017), pp. 419–435.

47. A. GUNAWAN, G. KENDALL, B. MCCOLLUM, H.-V. SEOW, AND L. S. LEE, *Vehicle routing: Review of benchmark datasets*, Journal of the Operational Research Society, 72 (2021), pp. 1794–1807.

48. A. GUTIÉRREZ-SÁNCHEZ AND L. B. ROCHA-MEDINA, *VRP variants applicable to collecting donations and similar problems: A taxonomic review*, Computers & Industrial Engineering, 164 (2022). Article 107887.

49. H. HAN AND E. PONCE-CUETO, *Waste collection vehicle routing problem: Literature review*, Promet - Traffic & Transportation, 27 (2015), pp. 345–358.

50. L. HEILIG AND S. VOSS, *Inter-terminal transportation: An annotated bibliography and research agenda*, Flexible Services and Manufacturing Journal, 29 (2017), pp. 35–63.

51. S. C. HO, W. Y. SZETO, Y.-H. KUO, J. M. Y. LEUNG, M. PETERING, AND T. W. H. TOU, *A survey of dial-a-ride problems: Literature review and recent developments*, Transportation Research Part B, 111 (2018), pp. 395–421.

52. M. IORI AND S. MARTELLO, *Routing problems with loading constraints*, TOP, 18 (2010), pp. 4–27.

53. M. IORI AND S. MARTELLO, *An annotated bibliography of combined routing and loading problems*, Yugoslav Journal of Operations Research, 23 (2013), pp. 311–326.

54. D. JAYARATHNA, G. LANEL, AND Z. JUMAN, *A contemporary recapitulation of major findings of vehicle routing problems: Models and methodologies*, International Journal of Recent Technology and Engineering, 8 (2019), pp. 581–585.

55. D. G. N. D. JAYARATHNA, G. H. J. LANEL, AND Z. A. M. S. JUMAN, *Survey on ten years of multi-depot vehicle routing problems: Mathematical models, solution methods and real-life applications*, Sustainable Development Research, 3 (2021), pp. 36–47.

56. D. J. N. D. JAYARATHNA, G. H. J. LANEL, Z. A. M. S. JUMAN, AND C. A. KANKANAMGE, *Modeling of an optimal outbound logistics system (a contemporary review study on effects of vehicle routing, facility location and locational routing problems)*, International Journal of Humanities and Social Science Invention, 8 (2019), pp. 8–30.

57. N. JAYARATHNA, J. LANEL, AND Z. A. M. S. JUMAN, *Five years of multi-depot vehicle routing problems*, Journal of Sustainable Development of Transport and Logistics, 5 (2020), pp. 109–123.

58. N. JOZEFOWIEZ, F. SEMET, AND E.-G. TALBI, *Multi-objective vehicle routing problems*, European Journal of Operational Research, 189 (2008), pp. 293–309.

59. S. KARAKATIČ AND V. PODGORELEC, *A survey of genetic algorithms for solving multi depot vehicle routing problem*, Applied Soft Computing, 27 (2015), pp. 519–532.

60. C. KOÇ, T. BEKTAŞ, O. JABALI, AND G. LAPORTE, *Thirty years of heterogeneous vehicle routing*, European Journal of Operational Research, 249 (2016), pp. 1–21.

61. C. KOÇ AND G. LAPORTE, *Vehicle routing with backhauls: Review and research perspectives*, Computers & Operations Research, 91 (2018), pp. 79–91.

62. C. KOÇ, G. LAPORTE, AND I. TÜKENMEZ, *A review of vehicle routing with simultaneous pickup and delivery*, Computers & Operations Research, 122 (2020). Article 104987.

63. G. D. KONSTANTAKOPOULOS, S. P. GAYIALIS, AND E. P. KECHAGIAS, *Vehicle routing problem and related algorithms for logistics distribution: A literature review and classification*, Operational Research, 22 (2022), pp. 2033–2062.

64. A. A. KOVACS, B. L. GOLDEN, R. F. HARTL, AND S. N. PARRAGH, *Vehicle routing problems in which consistency considerations are important: A survey*, Networks, 64 (2014), pp. 192–213.

65. S. KRITZINGER, K. F. DOERNER, F. TRICOIRE, AND R. F. HARTL, *Adaptive search techniques for problems in vehicle routing, part i: A survey*, Yugoslav Journal of Operations Research, 25 (2015), pp. 3–31.

66. S. KRITZINGER, K. F. DOERNER, F. TRICOIRE, AND R. F. HARTL, *Adaptive search techniques for problems in vehicle routing, part ii: A numerical comparison*, Yugoslav Journal of Operations Research, 25 (2015), pp. 169–184.

67. I. KUCUKOGLU, R. DEWIL, AND D. CATTRYSSE, *The electric vehicle routing problem and its variations: A literature review*, Computers & Industrial Engineering, 161 (2021). Article 107650.

68. R. LAHYANI, M. KHEMAKHEN, AND F. SEMET, *Rich vehicle routing problems: From a taxonomy to a definition*, European Journal of Operational Research, 241 (2015), pp. 1–14.

69. G. LAPORTE, *What you should know about the vehicle routing problem*, Naval Research Logistics, 54 (2007), pp. 811–819.

70. G. LAPORTE, *Fifty years of vehicle routing*, Transportation Science, 43 (2009), pp. 408–416.

71. Y.-C. LIANG, V. MINANDA, AND A. GUNAWAN, *Waste collection routing problem: A mini-review of recent heuristic approaches and applications*, Waste Management & Research, (2021), pp. 1–19.

72. Y.-J. LIANG AND Z.-X. LUO, *A survey of truck-drone routing problem: Literature review and research prospects*, Journal of the Operations Research Society of China, 10 (2022), pp. 343–377.

73. C. LIN, K. L. CHOY, G. T. S. HO, S. H. CHUNG, AND H. Y. LAM, *Survey of green vehicle routing problem: Past and future trends*, Expert Systems with Applications, 41 (2014), pp. 1118–1138.

74. C. Y. LIONG, I. WAN ROSMANIRA, O. KHAIRUDDIN, AND M. ZIROUR, *Vehicle routing problem: Models and solutions*, Journal of Quality Measurement and Analysis, 4 (2008), pp. 205–218.

75. R. B. LOPES, C. FERREIRA, B. S. SANTOS, AND S. BARRETO, *A taxonomical analysis, current methods, and objectives on location-routing problems*, International Transactions in Operational Research, 20 (2013), pp. 795–822.

76. G. MACRINA, L. DI PUGLIA PUGLIESE, F. GUERRIERO, AND G. LAPORTE, *Drone-aided routing: A literature review*, Transportation Research Part C, 120 (2020). Article 102762.

77. S. MARA, R. KUO, AND A. ASIH, *Location-routing problem: A classification of recent research*, International Transactions in Operational Research, 28 (2021), pp. 2941–2983.

78. Y. MARINAKIS AND A. MIGDALAS, *Annotated bibliography in vehicle routing*, Operational Research: An International Journal, 7 (2007), pp. 27–46.

79. E. MARREKCHI, W. BESBES, D. DHOUIB, AND E. DEMIR, *A review of recent advances in the operations research literature on the green routing problem and its variants*, Annals of Operations Research, 304 (2021), pp. 529–574.

80. T. A. S. MASUTTI AND L. N. DE CASTRO, *Bee-inspired algorithms applied to vehicle routing problems: A survey and a proposal*, Mathematical Problems in Engineering, 2017 (2017). Article 3046830.

81. P. MATL, R. F. HARTL, AND T. VIDAL, *Workload equity in vehicle routing problems: A survey and analysis*, Transportation Science, 52 (2018), pp. 239–260.

82. R. K. MAVI, M. GOH, N. K. MAVI, F. JIE, K. BROWN, S. BIERMANN, AND A. A. KHANFAR, *Cross-docking: A systematic literature review*, Sustainability, 12 (2020). Article 4789.

83. Q. MENG, S. WANG, H. ANDERSSON, AND K. THUN, *Containership routing and scheduling in liner shipping: Overview and future research directions*, Transportation Science, 48 (2014), pp. 265–280.

84. R. MOGHDANI, K. SALIMIFARD, E. DEMIR, AND A. BENYETTOU, *The green vehicle routing problem: A systematic literature review*, Journal of Cleaner Production, 279 (2021). Article 123691.

85. N. H. MOIN AND S. SALHI, *Inventory routing problems: A logistical overview*, Journal of the Operational Research Society, 58 (2007), pp. 1185–1194.

86. J. R. MONTOYA-TORRES, J. L. FRANCO, S. N. ISAZA, H. F. JIMÉNEZ, AND N. HERAZO-PADILLA, *A literature review on the vehicle routing problem with multiple depots*, Computers & Industrial Engineering, 79 (2015), pp. 115–129.

87. S. MOONS, K. RAMAEKERS, A. CARIS, AND Y. ARDA, *Integrating production scheduling and vehicle routing decisions at the operational decision level: A review and discussion*, Computers & Industrial Engineering, 104 (2017), pp. 224–245.

88. A. MOR AND M. G. SPERANZA, *Vehicle routing problems over time: A survey*, 4OR, 18 (2020), pp. 129–149.

89. M. C. MOURÃO AND L. S. PINTO, *An updated annotated bibliography on arc routing problems*, Networks, 70 (2017), pp. 144–194.

90. H. S. NA, S. J. KWEON, AND K. PARK, *Characterization and design for last mile logistics: A review of the state of the art and future directions*, Applied Sciences, 12 (2022). Article 118.

91. G. NAGY AND S. SALHI, *Location-routing: Issues, models and methods*, European Journal of Operational Research, 177 (2007), pp. 649–672.

92. N. M. E. NORMASARI AND N. LATHIFAH, *Heterogeneous fleet green vehicle routing problem: A literature review*, Angkasa Jurnal Ilmiah Bidang Teknologi, 13 (2021), pp. 49–57.

93. J. K. ODUM, A. OWUSU-ADDO, N. KYERE-SACRIFICE, AND A.-A. KWARTENG, *A review of disruption management in vehicle routing problem (vrp), transport design and scheduling*, International Journal of Multidisciplinary Studies and Innovative Research, 7 (2021), pp. 442–453.

94. B. H. OJEDA RIOS, E. C. XAVIER, F. K. MIYAZAWA, P. AMORIM, E. CURCIO, AND M. J. SANTOS, *Recent dynamic vehicle routing problems: A survey*, Computers & Industrial Engineering, 160 (2021). Article 107604.

95. M. OSTERMEIER, T. HENKE, A. HÜBNER, AND G. WÄSCHER, *Multi-compartment vehicle routing problems: State-of-the-art, modeling framework and future directions*, European Journal of Operational Research, (2021), pp. 799–817.

96. A. OTTO, N. AGATZ, J. CAMPBELL, B. GOLDEN, AND E. PESCH, *Optimization approaches for civil applications of unmanned aerial vehicles (uavs) or aerial drones: A survey*, Networks, 72 (2018), pp. 411–458.

97. J. OYOLA, H. ARNTZEN, AND D. L. WOODRUFF, *The stochastic vehicle routing problem, a literature review, part ii: Solution methods*, EURO Journal on Transportation and Logistics, 6 (2017), pp. 349–388.

98. J. OYOLA, H. ARNTZEN, AND D. L. WOODRUFF, *The stochastic vehicle routing problem, a literature review, part i: Models*, EURO Journal on Transportation and Logistics, 7 (2018), pp. 193–221.

99. D. C. PARASKEVOPOULOS, G. LAPORTE, P. P. REPOUSSIS, AND C. D. TARANTILIS, *Resource constrained routing and scheduling: Review and research prospects*, European Journal of Operational Research, 263 (2017), pp. 737–754.

100. J. PARK AND B.- I. KIM, *The school bus routing problem: A review*, European Journal of Operational Research, 202 (2010), pp. 311–319.

101. Y. PARK AND J. CHAE, *A review of the solution approaches used in recent g-vrp (green vehicle routing problem)*, International Journal of Advanced Logistics, 3 (2014), pp. 27–37.

102. S. PARRAGH, K. F. DOERNER, AND R. F. HARTL, *A survey on pickup and delivery problems part i: Transportation between customers and depot*, Journal fur Betriebswirtschaft, 58 (2008), pp. 21–51.

103. S. PARRAGH, K. F. DOERNER, AND R. F. HARTL, *A survey on pickup and delivery problems part ii: Transportation between pickup and delivery locations*, Journal fur Betriebswirtschaft, 58 (2008), pp. 81–117.

104. G. PEKER AND D. TÜRSEL ELIIYI, *Shuttle bus service routing: A systematic literature review*, Pamukkale Univ Muh Bilim Derg, 28 (2022), pp. 160–172.

105. V. PILLAC, M. GENDREAU, C. GUERET, AND A. L. MEDAGLIA, *A review of dynamic vehicle routing problems*, European Journal of Operational Research, 225 (2013), pp. 1–11.

106. H. POLLARIS, K. BRAEKERS, A. CARIS, G. JANSSENS, AND S. LIMBOURG, *Vehicle routing problems with loading constraints: State-of-the-art and future directions*, OR Spectrum, 37 (2015), pp. 297–330.

107. J.- Y. POTVIN, *State-of-the art review—evolutionary algorithms for vehicle routing*, INFORMS Journal on Computing, 21 (2009), pp. 518–548.

108. C. PRODHON AND C. PRINS, *A survey of recent research on location-routing problems*, European Journal of Operational Research, 238 (2014), pp. 1–17.

109. H. N. PSARAFTIS AND C. A. KONTOVAS, *Speed models for energy-efficient maritime transportation: A taxonomy and survey*, Transportation Research Part C: Emerging Technologies, 26 (2013), pp. 331–351.

110. H. N. PSARAFTIS, M. WEN, AND C. A. KONTOVAS, *Dynamic vehicle routing problems: Three decades and counting*, Networks, 67 (2016), pp. 3–31.

111. H. QIN, X. SU, T. REN, AND Z. LUO, *A review on the electric vehicle routing problems: Variants and algorithms*, Frontiers of Engineering Management, 8 (2021), pp. 370–389.

112. N. RINCON- GARCIA, B. J. WATERSON, AND T. J. CHERRETT, *Requirements from vehicle routing software: Perspecitves from literature, developers and the freight industry*, Transport Reviews, 38 (2018), pp. 117–138.

113. U. RITZINGER, J. PUNCHINGER, AND R. F. HARTL, *A survey on dynamic and stochastic vehicle routing problems*, International Journal of Production Research, 54 (2016), pp. 215–231.

114. D. ROJAS VILORIA, E. L. SOLANO-CHARRIS, A. MUÑOZ VILLAMIZAR, AND J. R. MONTOYA-TORRES, *Unmanned aerial vehicles/drones in vehicle routing problems: A literature review*, International Transactions in Operational Research, (2020), pp. 1–32.

115. R. F. ROLDÁN, R. BASAGOITI, AND L. C. COELHO, *A survey on the inventory-routing problem with stochastic lead times and demands*, Journal of Applied Logic, 24 (2017), pp. 15–24.

116. D. G. ROSSIT, D. VIGO, F. TOHME, AND M. FRUTOS, *Visual attractiveness in routing problems: A review*, Computers & Operations Research, 103 (2019), pp. 13–34.

117. R. V. SANGEETHA AND A. G. SRINIVASAN, *A review of static, dynamic and stochastic vehicle routing problems in home healthcare*, European Journal of Molecular & Clinical Medicine, 7 (2020), pp. 5037–5046.

118. H. SHAABANI, *A literature review of the perishable inventory routing problem*, The Asian Journal of Shipping and Logistics, 38 (2022), pp. 143–161.

119. N. SOEFFKER, M. W. ULMER, AND D. C. MATTFELD, *Stochastic dynamic vehicle routing in the light of prescriptive analytics: A review*, European Journal of Operational Research, 298 (2022), pp. 801–820.

120. S. SRIVATSA SRINIVAS AND M. S. GAJANAND, *Vehicle routing problem and driver behaviour: A review and framework for analysis*, Transport Reviews, 37 (2017), pp. 590–611.

121. Y. SUN, S. KIRTONIA, AND Z.-L. CHEN, *A survey of finished vehicle distribution and related problems from an optimization perspective*, Transportation Research Part E, 149 (2021). Article 102302.

122. S.-Y. TAN AND W.-C. YEH, *The vehicle routing problem: State-of-the-art classification and review*, Applied Sciences, 11 (2021). Article 10295.

123. A. THIBBOTUWAWA, G. BOCEWICZ, P. NIELSEN, AND Z. BANASZAK, *Unmanned aerial vehicle routing problems: A literature review*, Applied Sciences, 10 (2020). Article 4504.

124. E. TORO, A. ESCOBAR, AND M. GRANADA, *Literature review on the vehicle routing problem in the green transportation context*, Luna Azul, 42 (2016), pp. 362–387.

125. M. W. ULMER, J. C. GOODSON, D. C. MATTFELD, AND B. W. THOMAS, *On modeling stochastic dynamic vehicle routing problems*, EURO Journal on Transportation and Logistics, 9 (2020). Article 100008.

126. D. M. UTAMA, S. K. DEWI, A. WAHID, AND I. SANTOSO, *The vehicle routing problem for perishable goods: A systematic review*, Cogent Engineering, 7 (2020). Article 1816148.

127. A. UTAMIMA AND A. DJUNAIDY, *Agricultural routing planning: A narrative review of literature*, Procedia Computer Science, 197 (2022), pp. 693–700.

128. P. VANSTEENWEGEN, W. SOUFFRIAU, AND D. VAN OUDHEUSDEN, *The orienteering problem: A survey*, European Journal of Operational Research, 209 (2011), pp. 1–10.

129. T. VIDAL, T. G. CRAINIC, M. GENDREAU, AND C. PRINS, *Heuristics for multi-attribute vehicle routing problems: A survey and synthesis*, European Journal of Operational Research, 231 (2013), pp. 1–21.

130. T. VIDAL, G. LAPORTE, AND P. MATL, *A concise guide to existing and emerging vehicle routing problem variants*, European Journal of Operational Research, 286 (2020), pp. 401–416.

131. X. WANG AND E. WASIL, *On the road to better routes: Five decades of published research on the vehicle routing problem*, Networks, 77 (2021), pp. 66–87.

132. Y. XIAO, Y. ZHANG, I. KAKU, R. KANG, AND X. PAN, *Electric vehicle routing problem: A systematic review and a new comprehensive model with nonlinear energy recharging and consumption*, Renewable and Sustainable Energy Reviews, 151 (2021). Article 111567.

133. I. ZENNARO, S. FINCO, M. CALZAVARA, AND A. PERSONA, *Implementing e-commerce from logistic perspective: Literature review and methodological framework*, Sustainability, 14 (2022). Article 911.

134. H. ZHANG, H. GE, J. YANG, AND Y. TONG, *Review of vehicle routing problems: Models, classification and solving algorithms*, Archives of Computational Methods in Engineering, 29 (2022), pp. 195–221.

135. T. P. V. ZIS, H. N. PSARAFTIS, AND L. DING, *Ship weather routing: A taxonomy and survey*, Ocean Engineering, 213 (2020). Article 107697.

Addendum

Since we have completed our compilation of survey articles in the field of vehicle routing, new articles have appeared and we found one article from 2010 that we missed. In order to be as complete and up-to-date as possible, we list these additional articles below in alphabetical order with respect to the authors:

136. G. Berbeglia, J-F. Cordeau, and G. Laporte, *Dynamic pickup and delivery problems*, European Journal of Operational Research, 202 (2010), pp. 8–15.
137. M. Diah, A. Setyanto, and E.T. Luthfi, *Systematic literature review of particle swarm optimization implementation for time-dependent vehicle routing problem*, Journal Online Informatika, 7 (2022), pp. 38–45.
138. H. Li, J. Chen, F. Wang, and M. Bai, *Ground-vehicle and unmanned-aerial-vehicle routing problems from two-echelon scheme perspective: A review*, European Journal of Operational Research, 294 (2021), pp. 1078–1095.
139. A. Nura and S. Abdullahi, *A systematic review of multi-depot vehicle routing problems*, Systematic Literature Review and Meta-Analysis Journal, 3 (2022), pp. 51–60.
140. S. Srinivas, S. Ramachandiran, and S. Rajendran, *Autonomous robot-driven deliveries: A review of recent developments and future directions*, Transportation Research Part E, 165 (2022). Article 102834.
141. C. Ye, W. He, and H. Chen, *Electric vehicle routing models and solution algorithms in logistics distribution: A systematic review*, Environment Science and Pollution Research, (2022), https://doi.org/10.1007/s11356-022-21559-2.

B. Golden et al., *The Evolution of the Vehicle Routing Problem*, Synthesis Lectures on Operations Research and Applications,
https://doi.org/10.1007/978-3-031-18716-2

Printed in the United States
by Baker & Taylor Publisher Services